Monographs of the Physiological Society No. 40

Diving and asphyxia

A comparative study of animals and man

DIVING AND ASPHYXIA

A comparative study of animals and man

ROBERT ELSNER

Professor of Marine Science, University of Alaska

BRETT GOODEN

Adelaide, South Australia
Formerly Lecturer in Physiology, University of Nottingham

CAMBRIDGE UNIVERSITY PRESS

Cambridge
London New York New Rochelle
Melbourne Sydney

Published by the Press Syndicate of the University of Cambridge
The Pitt Building, Trumpington Street, Cambridge CB2 1RP
32 East 57th Street, New York, NY 10022, USA
296 Beaconsfield Parade, Middle Park, Melbourne 3206, Australia

First published 1983

Printed in Great Britain at the University Press, Cambridge

Library of Congress catalogue card number: 82-21998

British Library cataloguing in publication data
Elsner, Robert
Diving and asphyxia. – (Monographs of the
Physiological Society; 40)
1. Diving, Submarine – Physiological aspects
2. Underwater physiology
I. Title II. Gooden, Brett
III. Series
612'.04   RC1015

ISBN 0 521 25068 4

SE

Monographs of the Physiological Society

v

Volumes marked with an asterisk are now out of print.

CONTENTS

Contents

PREFACE

We have assembled in this monograph our views regarding the physiological events that are collectively described as the *diving response*. The emphasis is mostly but not exclusively mammalian, as is appropriate to our experience. Adaptations to diving asphyxia were originally identified only in animals that live in aquatic habitats. It has since become evident that the natural divers display a well-developed variation of a more general defence against asphyxia. Much can be learned about asphyxia and related cardiovascular and metabolic responses from field and laboratory studies of diving animals, the natural specialists whose reactions yield conveniently to experimental manipulation. An even broader perspective suggests that the species so tolerant of diving asphyxia are part of a continuum of animals which respond to hostile environments and disturbances of homeostasis by strategic retreat into conditions of depressed metabolism. Knowledge derived from the study of diving animals contributes to biological understanding and also holds promise for practical applications to clinical medicine.

We are much in debt to colleagues and students for suggestions and discussions, but our failures in explanation and judgement are our own responsibility. We owe special and personal gratitude to those two pioneers in this field, Laurence Irving and P. F. Scholander, both recently deceased, who set high standards some forty years ago for deciphering the mysteries governing the adaptations of diving mammals and birds. Whatever good comes of this publication is dedicated to them. We are very grateful to Professor M. de Burgh Daly, colleague and friend, for stimulating discussions and support and for contributions to our thinking about control mechanisms and medical implications.

Preface

Two anonymous reviewers suggested important changes which substantially clarified and improved the text. Our wives, Elizabeth Elsner and Lesley Gooden, willingly read and re-read parts of the manuscript and made useful suggestions for its improvement. Mrs Helen Stockholm, Publications Supervisor, and typists Suzette Carlson, Martha Fisk, Roseanne Lamoreaux, Mauricette Nicpon, Nancy Ricci and Tricia Witmer, all at the Institute of Marine Science, University of Alaska, remained helpful and tolerant throughout the turmoil of producing the manuscript. Mrs Ana Lea Vincent drafted some of the figures.

Our experimental subjects, marine mammals, human breath-holders and others, cooperated by revealing some of their physiological secrets. The research was supported by the US Public Health Service National Heart, Lung and Blood Institute, the National Science Foundation, the US Antarctic Research Program, the Alaska Heart Association, the National Health and Medical Research Council (Australia), the National Heart Foundation of Australia, the Fulbright–Hays Programme, the Australian–American Educational Foundation, the Moody Foundation, Sea World, Seward Fisheries, the Universities of Adelaide, Texas, Nottingham, Alaska and California (Scripps Institution of Oceanography) and the Royal Society and Wellcome Trust (to Professor M. de Burgh Daly). Professors R. F. Whelan and A. D. M. Greenfield encouraged and supported our joint enterprise.

Our families contributed to the success of our research by helping with the care of experimenters and experimental animals: sea lions, seals, dolphins, a pilot whale, echidnas, dogs and new-born lambs. Dedicated technicians, especially Mr James Wright and Mrs Sally Dunker, kept their composure and resolve despite demands which were sometimes extraordinary and bizarre.

Robert Elsner
Fairbanks, Alaska
Brett A. Gooden
Adelaide, South Australia

1 THE BIOLOGICAL SETTING

The diving mammals offer the physiologist a natural
experiment which shows how long and by means of what
respiratory adjustment a mammal can endure asphyxia
(Laurence Irving, 1939).

Depriving living organisms of oxygen and of the opportunity
to rid themselves of carbon dioxide leads inevitably to
progressive disruption of cellular biochemical processes,
physiological dysfunction and the undoing of biological
integrations which are collectively essential for homeostatic
life. This condition, asphyxia, is a threat which lurks never far
away from living vertebrate organisms. The so-called lower
animals are generally more resistant to it than the higher,
more complex forms which, for that complexity, are dependent upon a nearly continuous respiratory exchange.
Asphyxia, and responses to it, can take several forms. It can
characterize the disturbance of blocked respiration in a whole
animal and it can take place in a regional tissue which is
deprived of the perfusing blood required for the exchanges
that are necessary for cellular survival. It is what the diving
seal or duck experiences when it holds its breath and
submerges in water. That exposure is usually brief and non-
threatening, but if asphyxia continues without interruption,
it leads inevitably to cellular disruption and organism death.
In the considerations which follow, examples will be drawn
from phenomena other than diving, and from both aquatic and
terrestrial species, including man.

Resistance to asphyxia varies with species, age and special
adaptation. Tolerance to ischaemia, the occlusion of blood
flow, also varies within the animal according to the type of
cells in the tissue deprived of circulation. Just as the time
during which an animal can hold its breath and remain under
water varies from species to species, so does the tolerance of

1

tissues to ischaemia vary from organ to organ. The harbour seal can remain under water for a maximum diving time of 25 min. The Weddell seal of Antarctica can remain submerged for at least 60 min. Some ducks can dive for 10–15 min. Humans, with some practice, can breath-hold under water for about 2 min.

Similarly, a hierarchy exists among the tissues of the body. For example, the tissues in human limbs can be deprived of blood for 30 min or more without damage. On the other hand, it is obvious that the central nervous system, more specifically those portions of it involved in maintenance of consciousness, cannot sustain activity for more than a few seconds without a continuous supply of blood rich in oxygen. The potentially disruptive effects in these examples result from interruption in the supply of oxygen and its subsequent decline in the tissues, the accumulation of carbon dioxide and of other metabolic products, and the increasing acidity. The pathophysiological consequences of insufficient oxygen were eloquently discussed by Barcroft (1920) in a landmark article.

Pure hypoxia, that is, oxygen depletion with normal blood and tissue concentrations of carbon dioxide and hydrogen ions, is rare in nature. Acute altitude exposure, for instance, usually results in hypoxia together with hypocapnia and alkalosis produced by an associated increase in pulmonary ventilation. But asphyxial interference with respiration, as in breath-holding, diving, suffocation and loss of consciousness, leads to hypoxia, hypercapnia and acidosis, all of which increase progressively and relentlessly as asphyxia develops.

Adaptations which permit seals, ducks and other naturally diving animals to tolerate long submersions provide good examples of physiological and biochemical specializations. However, as they are understood today, these adaptations suggest no fundamentally new or previously unrecorded mechanisms. They are qualitatively related to reactions and regulatory mechanisms common to vertebrates, but species differ in their quantitative development. They can perhaps be best considered as extensions of the normal responses which guarantee stable conditions of life. In more general terms, asphyxial adaptations apply to invertebrates as well.

The train of events that takes place as seals dive demonstrates these adaptations. Seals of the family Phocidae, capable of long dives, have special capabilities for oxygen storage. Their circulating blood volume is elevated, and their blood is richer in circulating haemoglobin than that of most other mammals (Lenfant, 1969). Further oxygen storage is accomplished by the high myoglobin contents in skeletal muscles (Robinson, 1939; Scholander, Irving & Grinnell, 1942*a*; Blessing & Hartschen-Niemeyer, 1969). Their lung volumes are not unusually different from those of terrestrial mammals, and some species usually exhale before diving. Thus, the major oxygen source which is readily available to central organs is stored within the blood itself. Accordingly, the total oxygen-storage capacities of harbour seal and Weddell seal blood, normalized for differences in body weight, are approximately two and three times, respectively, that of man (Lenfant *et al.*, 1969*b*).

The resting metabolic rates of seals are somewhat higher than those of comparably sized terrestrial mammals, some being approximately twice as high, while body temperatures are similar to those of land mammals. The elevated metabolic rate and thick subcutaneous blubber (representing 25–50% of body weight) provide the heat production and insulation necessary for maintenance of normal internal temperature in the cooling water environment (reviewed by Irving, 1969, Elsner *et al.*, 1977*b* and Blix & Steen, 1979). It can be readily calculated that if the submerged seal were to continue metabolizing at the same rate as before diving, its oxygen storage, despite its magnitude, would not be sufficient to maintain that rate during long dives. Clearly, some other responses must be essential for its survival.

The mammalian species which are adapted to aquatic habitats range in size from the blue whale to the water shrew, thus including the biggest animal ever to have lived and one of the smallest. Their maximum diving times vary from less than 1 min to 1- to 2-h submersions of sperm and bottlenosed whales (Irving, 1939). Some turtles can remain underwater for days on end (Belkin, 1963; Robin *et al.*, 1964). Weddell seals can dive to 600 m, sperm whales to more than 1000 m. Pathological accumulation of inert gas (decompression sickness), which might occur in repetitive diving, is avoided by

3

virtue of structural adaptations which allow for thoracic compression and isolation of pulmonary air from perfusing blood during dives. These include a flexible rib cage, stiffened alveolar ducts and attachments of the diaphragm such as to permit some shifting of abdominal contents into the thorax (Scholander, 1940; Kooyman, 1972; Kooyman *et al.*, 1972).

Natural observations and laboratory experiments

There has been a tendency through the years to look upon the reactions of aquatic species, such as seals and ducks, to diving experiments as constituting a fixed set of physiological events. The dogma suggests that all examples of diving, whether occurring in nature or in the laboratory, result in similar responses. Cessation of breathing, slowing of the heart rate and changes in the distribution of the circulating blood to favour the most vital organs have become fixed concepts to an extent greater than is warranted by the results of critical studies extending over several years. That view of diving physiology is not the whole story; rather, the responses vary in timing and intensity, just as do most biological phenomena.

In his classical work of 1940, Scholander pointed out that a seal freely submerging in water sometimes failed to show diving bradycardia. However, seals reacted to startle and loud noise with sudden heart rate slowing. Experimental diving studies in which animals were trained to dive upon a signal resulted in variable responses in the sea lion (Elsner, Franklin & Van Citters, 1964*a*), the harbour seal (Elsner, 1965; Elsner *et al.*, 1966*a*) and the dolphin (Elsner, Kenney & Burgess, 1966*b*). There was usually less decline in heart rate during trained immersion than during forced immersion (sea lion and harbour seal), but in dolphins the bradycardia was intensified during trained dives. The generalization which appears to apply is that some major component of the diving response is determined by the intention, conditioning or psychological perspective of the animal being studied. Thus, the diver acts as though it produces the most intense diving response when the need for achieving maximum diving duration is anticipated.

4

Perhaps trained dolphins are more highly motivated than trained seals.

Kooyman *et al* (1980) described dives of up to 20 min in Weddell seals (which are capable of maximum dives exceeding 60 min) in which little or no dependence upon the anaerobic resources required for longer dives was made, although cardiovascular changes, signalled by bradycardia, presumably occurred (Kooyman & Campbell, 1972). Oxygen reserves were apparently sufficient to allow the dives to be made aerobically. If, as seems likely, these reserves are principally contained within the blood and oxygenated myoglobin of skeletal muscle, then the reactions of the voluntarily diving Weddell seals resembled those of restrained diving harbour seals in a previous study in which myoglobin oxygen was rapidly depleted before lactate production commenced in about 10 min, earlier in seals which struggled (Scholander *et al.*, 1942*a*: Fig. 1.1). One is struck by the similarity of the responses in the two examples, rather than by differences. Authors of both studies remarked upon the observation that the preponderance of free dives in nature are brief and aerobic, indicating that anaerobic responses, which are more exhausting and require longer recovery, are seldom brought into operation. Some, perhaps most, marine mammals are able, by virtue of efficient pulmonary gas exchange, to recover promptly from such dives, replenish oxygen stores and prepare for the next dive (Olsen, Hale & Elsner, 1969; Denison & Kooyman, 1973).

Butler & Woakes (1979) and Kanwisher, Gabrielsen & Kanwisher (1981) showed that freely diving birds make many short dives without changes in heart rate. Clearly, most of the dives performed by aquatic animals in nature are of short duration, and they apparently depend largely upon oxygen-sustained rather than anaerobic mechanisms. Ducks and seals doubtless find it uncomfortable and exhausting to push their diving habit to the anaerobic extreme of their reserve capabilities. Although that physiological reserve is rarely invoked, it is, nevertheless, a resource upon which survival may depend. From these considerations there emerges the concept of graded responses to diving, varying in intensity depending upon severity of the imposed stress and of the

5

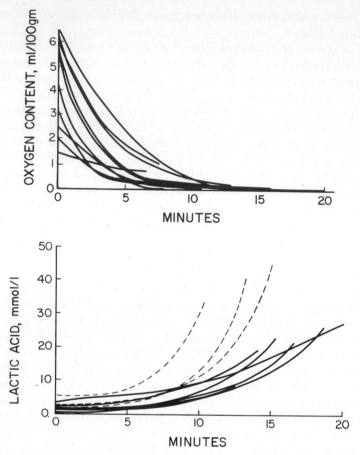

Fig. 1.1. Depletion of myoglobin-bound oxygen (above) and increase in lactic acid concentration (below) in skeletal muscle of harbour seals *Phoca vitulina* during restrained dives. The solid lines represent dives of individual seals; the dashed lines represent dives during which the seals struggled against the restraint. (Redrawn from Scholander *et al.*, 1942*a*.)

degree to which control is exercised by higher cortical functions.

The understanding of diving biology is enhanced by examining the natural history of diving species. It is equally important that we study the full range of physiological responses of which these animals are capable by submitting them to experimental dives. Controlled laboratory investigations require either the use of trained animals or, more drastically, the interventions of restraint and anaesthesia. These approaches have all been usefully employed in various studies. Some success has been achieved in combining laboratory and fieldwork in natural settings.

A major response to most diving situations, natural and experimental, is the selective redistribution of the circulation. Inasmuch as many of the tissues of the body can tolerate oxygen deprivation for longer than the more vital ones, the heart and the brain, the available blood oxygen could be conserved by its preferential distribution to those vital organs for their immediate needs while the remaining organs, deprived of circulation, survive on anaerobic metabolic processes or, in the case of skeletal muscle, on oxygen bound to myoglobin. Modern understanding of these diving responses rests on the foundation of classical research reports by Irving (1934, 1939) and Scholander (1940) and their later collaborative publications. Some aspects of these adaptive mechanisms exist in many species which are not habitual divers when they are exposed to asphyxia. (General reviews: Scholander, 1962, 1963, 1964; Irving, 1964; Andersen, 1966, 1969; Elsner *et al.*, 1966*a*; Robin, 1966; Harrison & Kooyman, 1968; Strauss, 1970; Jones & Johansen, 1972; Ridgway, 1972; Galantsev, 1977; Kooyman, Castellini & Davis, 1981; Butler & Jones, 1982; Blix & Folkow, 1983).

The term 'diving response' refers to a sequence and collection of physiological events, including apnoea, bradycardia and redistribution of cardiac output, which are all under the control of multiple reflexes rather than forming a single 'diving reflex'. This latter term, which enjoys some popularity, conveys the incorrect impression that the events seen in diving constitute one reflex rather than many reflexes functioning together through their interactions.

Hypometabolism, a strategic retreat

Restricting the circulation of blood to a tissue or an organ results in a steady decline of the oxygen available for support of oxidative metabolic processes and subsequent dependence

7

upon whatever anaerobic resources are available. Eventually, a depression of metabolism takes place. This lowering of the rate at which the many complex chemical processes can occur, thus conserving metabolic energy, is, in fact, a central feature of the adaptations to asphyxia. Viewed in this context, the diving response represents but one specialized example of a widespread and general response of many living animals, invertebrate as well as vertebrate, to life-threatening situations. There are abundant examples in nature.

The phenomenon of dormancy, which is characterized by a lowering of metabolic activity, is widespread in nature. Marine invertebrates such as intertidal molluscs, which require submersion in water to obtain an optimum gas exchange, are regularly subjected to cyclic tidal conditions which leave them high and dry. They retreat from those threatening situations into a state of lowered metabolism, thus decreasing the need for immediate dependence upon life-supporting functions (Newell, 1973). Fish removed from water show the reduced and redistributed circulation typical of the diving response (Leivestad, Andersen & Scholander, 1957; Garey, 1962; Scholander, Bradstreet & Garey, 1962*b*). The marine invertebrate *Aplysia* responded to air exposure with bradycardia (Feinstein *et al.*, 1977). A warm-water species, *A. brasiliana*, showed a 43% average decrease in heart rate (Fig. 1.2), whereas a cold-water species, *A. californica*, showed only a 16.5% decrease. The bradycardia response in the warm-water species was shown to be largely neurally controlled via the abdominal ganglion. *Aplysia* are occasionally stranded on beaches or in tidal pools and are sometimes able to survive this asphyxial stress until they are once again submerged in the next tide. Hibernators and aestivators likewise retreat from the threats of decreased food supplies and thermal conflicts by lowering their metabolic rates. The highly active hummingbirds and bats experience nocturnal and daytime declines, respectively, in metabolic rate which provide them with an economical means of avoiding the necessity for obtaining a continuous food supply.

Snails react to dehydration by sealing off their shell openings and entering a dormant state (Schmidt-Nielsen,

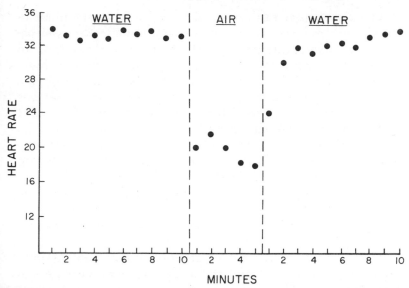

Fig. 1.2. Heart rate (beats/min) of *Aplysia brasiliana* during removal from water, which induces asphyxia. (Redrawn from Feinstein *et al.*, 1977.)

Taylor & Shkolnik, 1971). Desert frogs respond to threatened dehydration by torpor. They can remain sealed off, buried in the ground, awaiting the next rainfall, which may be months or years away. The African lungfish *Protopterus* similarly lies dormant in dried mud-holes during long dry periods (Smith, 1930). Diapause is a well-known dormant condition in insects (Lees, 1955).

In addition to those examples that have been mentioned here, it is now appreciated that conditions of lowered metabolic activity occur in a great variety of mammalian, bird and invertebrate species, culminating perhaps in the phenomenon of cryptobiosis, a death-like state entered into by several species of primitive animals when they become thoroughly dehydrated. They can remain in this depressed state for years and can then be revived by rehydration. Certain species of rotifers and nematodes are examples (Crowe & Cooper, 1971). Scholander *et al.* (1953) described similar metabolic responses in frozen arctic plants and animals.

9

The biological setting

The spectacular physiological events which take place in the process of mammalian birth provide yet another example of the implementation of a set of responses designed to protect the integrity of the central nervous system during a threat of asphyxia. The infant animal is exposed to asphyxia sometimes by disruption of placental blood flow and gas exchange, sometimes by compression of the umbilical cord and during the interval from the moment when it loses its gaseous lifeline to the maternal placental circulation until the moment when it first inflates its lungs. Its protection during these critical minutes depends upon adaptations for just that brief once-in-a-lifetime event. These include its anaerobic reserves and cardiovascular adjustments that favour its central nervous system at the expense of less sensitive tissues (Chapter 6).

An example of human hypometabolic retreat from disturbing environments is demonstrated by certain practitioners of Yoga in India. This voluntary reduction of metabolic rate was first documented by Anand, Chhina and Singh (1961) who studied a Yogi subject enclosed within an air-tight box. The investigation was inspired by reports of experienced Yogis who commit themselves to burial for several days in small subterranean chambers in order to achieve deep meditation in the absence of sensory stimulation. The burial chambers are most likely not air-tight, and recent evidence shows that the metabolic effects can be produced by the meditating Yogi without recourse to burial or enclosure. R. Elsner and H. C. Heller (unpublished results) studied a well-practised Yogi during 4 h of meditation in which his basal oxygen consumption was reduced by 45 % (Fig. 1.3). Brief hypometabolic states have been described by Wallace, Benson & Wilson (1971) in subjects trained in Transcendental Meditation.

The most remarkable hypometabolic states in mammals are displayed by hibernators. They attain low metabolic rates as a consequence of lowering the body temperature, and this involves a resetting of the hypothalamic regulator of body temperature. The threshold temperature for thermoregulatory responses is much reduced, as is the gain of the regulator (Heller & Hammel, 1972; Heller & Colliver, 1974). Humans and other mammals usually experience slightly

10

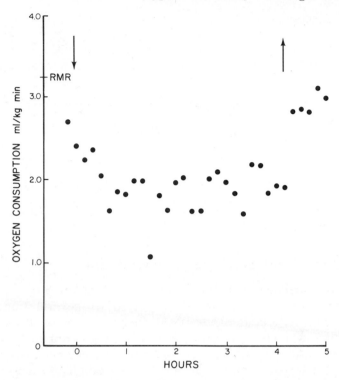

Fig. 1.3. Oxygen consumption of a Yogi during deep mediation, beginning and ending indicated by arrows. RMR, resting metabolic rate. (R. Elsner & H. C. Heller, unpublished results.)

reduced body temperatures and decreased metabolic rates during sleep. Physiological evidence suggests that the states of sleep and hibernation represent a continuum of the same basic processes involving progressive alterations in metabolism and temperature regulation (South *et al.*, 1969; Heller & Glotzbach, 1977; Walker *et al.*, 1977). Huntley *et al.* (1981) described the unusual conservation of metabolic energy by elephant seal pups during intermittent sleep. The overall reduction in oxygen consumption amounted to an average of 48% of the basal metabolic rate. Complete cessation of respiration occurred while the animals slept. The average duration of sleep apnoea was 10 min, the longest being 24 min.

11

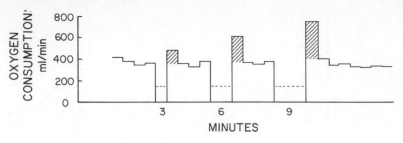

Fig. 1.4. Oxygen consumption of a grey seal *Halichoerus gryphus* measured before and after dives of 3, 6 and 9 min duration. Measurements were obtained during 3-min periods. The post-dive increase in oxygen consumption (cross-hatching, area equal to the area below the preceding dashed line) is less than predicted by the pre-dive level of consumption had diving not occurred. (Redrawn from Scholander, 1940.)

Reduced metabolic rate during diving

The occurrence of reduced metabolic rates during experimental diving asphyxia is supported by several lines of evidence. Scholander (1940) was the first to demonstrate this reaction when he found that the extra oxygen consumption of seals and ducks during recovery from quiet dives could not account for the amount predicted if metabolism during the dive was maintained at the pre-dive level (Fig. 1.4). Body temperatures decreased in diving ducks (Andersen, 1959) and seals along with the decrease in metabolism (Scholander, Irving & Grinnell, 1942*b*; Elsner, Hammel & Heller, 1975; Hammel *et al.*, 1977; Kooyman *et al.*, 1980). Therefore, they could be expected to repay only part of the accumulated oxygen deficit during recovery from the hypoxic diving episode (Dejours, 1975). However, the post-dive oxygen consumption of muskrats did not suggest reduced metabolic rate in that species during dives (Fairbanks & Kilgore, 1978).

Depression of metabolic rate was seen in diving ducks (Andersen, 1959) and alligators (Andersen, 1961). A calorimetric method showed a decline of sometimes as much as 90% in the metabolic rate of diving ducks (Pickwell, 1968). Direct calorimetry has shown similar responses in toads (Leivestad, 1960) and turtles (Jackson & Schmidt-Nielsen, 1966; Jackson, 1968). Heat production in turtles diving for

12

2–4 h fell to 15–20% of the pre-dive values. Pre-dive ventilation of freshwater turtles *Pseudemys scripta* with pure oxygen delayed the onset of reduced metabolic rate, suggesting that depressed metabolism was not triggered by reflexes associated with the act of diving but rather by mechanisms related to the reduction of available oxygen. When oxygen reserves were depleted in experimental dives, muscular activity steadily decreased and heat production reached its lowest level.

Diving turtles also developed slow heart rates and circulatory redistribution of cardiac output which was drastically lowered during diving and largely bypassed the polmonary circuit (White & Ross, 1966). Freshwater turtles survived for several hours in a nitrogen atmosphere (Johlin & Moreland, 1933; Belkin, 1963; Millen *et al.*, 1964; Robin *et al.*, 1964), and they could not tolerate long diving after injection of the glycolytic poison iodoacetate, which blocks anaerobic metabolism. Both observations testify to dependence upon anaerobic glycolytic processes for survival during long periods of anoxia (Belkin, 1962). However, even turtles are smart enough not to rely more than they must on their anaerobic reserves, and most free dives are aerobic (Ackerman & White, 1979). Remarkable anoxic tolerance has also been described in marine turtles (Lutz *et al.*, 1980). Marine iguanas of the Galapagos Islands depend upon anaerobic glycolysis for much of their diving capability (Bartholomew, Bennett & Dawson, 1976). Voluntarily diving monitor lizards consumed oxygen from the blood stores, at first rapidly and then more slowly, finally surfacing when arterial oxygen tension fell to about 30 mmHg (Wood & Johansen, 1974).

Recognition that such a wide range of species, from *Aplysia* to human beings, all show the ability to make a strategic retreat from environmental threats into states of metabolic depression, suggests the fundamental nature of this biological reaction. Cardiovascular adaptations leading to a redistribution of the blood flow that favours the more vital structures, thus accomplishing an overall sparing of metabolic energy, play a major role in many of the examples.

13

METABOLIC CONSERVATION BY CARDIOVASCULAR ADJUSTMENTS

Le coeur qui battait avant l'immersion environ 100 fois par minute, ne donne plus que 14 battements réguliers et profonds (Paul Bert, 1870).

Historical introduction

Robert Boyle (1670) was probably the first to experiment with diving animals when he weighted the feet of a duck and placed it in a container of water in an attempt to determine the tolerance of the duck to submersion. In 1870, the French physiologist Paul Bert made the important discovery that ducks experimentally submerged in water experienced a dramatic slowing of their heart rates. This observation initiated a long and still continuing series of investigations. Physiologists continue to reveal the intricate nature of the special diving abilities of aquatic species of birds, mammals and reptiles. In the process some interesting and unexpected discoveries in terrestrial animals, including man, have come to light.

Bert compared the duck and hen with regard to resistance to diving asphyxia. Bert suggested that the superior performance of ducks compared to hens was related to its larger blood volume to body weight ratio (1:17 in the duck, 1:30 in the hen), noting that bleeding the duck reduced its diving capability. Bert concluded that the advantage possessed by the duck could be accounted for by the greater oxygen storage capacity of its larger blood volume. This has turned out to be only part of the story.

Richet (1894, 1899) showed that ducks which had been bled to a blood volume to body weight ratio equivalent to or even less than that of hens still survived immersion for longer than might be expected on the basis of differences in relative blood volume alone. Furthermore, an important reduction in metabolic rate during diving was deduced on the basis that oxygen stores alone were insufficient to last throughout the dive (Langlois & Richet, 1898).

Despite the fact that oxygen storage is elevated in some species of diving animals, several investigators have recognized the disparity between the magnitude of those stores and the oxygen requirements that would be anticipated if metabolism proceeded at an undiminished rate during diving. Irving *et al.* (1935) suggested that oxygen might be conserved for vital organs that are essential to survival and which require a continuous source of oxygen, presumably the heart and the brain.

Substantial evidence, due principally to the pioneer research of Irving and Scholander, appeared in support of the concept of oxygen-sparing mechanisms that are largely dependent upon circulatory redistribution. The special adaptations of diving animals were viewed as variations in the general vertebrate framework rather than as differences in the qualitative nature of the responses. Blocking of the trachea resulted in decreased muscle blood flow with maintained or increased blood flow in the brain of the beaver, muskrats, cats, dogs and rabbits (Irving, 1938). Lactic acid produced by anaerobic metabolism increased only very little in the circulating arterial blood of restrained diving seals; more appeared late in a long dive and copious amounts were seen immediately after surfacing and the resumption of breathing (Fig. 2.1). This discovery was interpreted as showing that muscle perfusion was reduced during dives and that lactic acid which accumulated in those tissues was flushed out after diving by the re-established circulation (Scholander, 1940). The heart rate of the seal fell dramatically, often within one or two seconds of the start of the dive, to levels of intense bradycardia amounting to about one-tenth of the pre-dive rate. While these observations were amply demonstrated in grey seals *Halichoerus gryphus* and hooded seals *Cystophora cristata*, they were also shown to occur in varying and lesser degrees in ducks, penguins, rats and rabbits.

Scholander *et al.* (1942a) obtained further evidence for the lack of communication between circulating blood and muscle

15

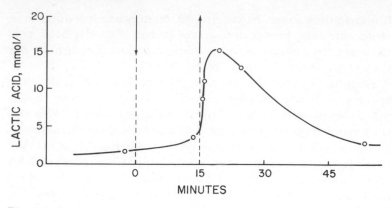

Fig. 2.1. Lactic acid concentration in the arterial blood of an experiment-ally dived grey seal *Halichoerus gryphus*. Dive indicated by arrows. (Redrawn from Scholander, 1940.)

tissue through the observation that the oxygen in myoglobin-rich muscle was depleted during the dive more rapidly than the oxygen in circulating arterial blood (Chapters 1 and 3). Since the affinity of myoglobin for oxygen is much greater than that of haemoglobin, circulatory isolation of the metabolizing muscles seems obviously beneficial. Thus, if the muscle circulation were not checked during diving, large quantities of oxygen would be removed from the blood by myoglobin. The oxygen stored in myoglobin is presumably sufficient for several minutes' diving with little muscular activity by the submerged seal: up to 10 min in harbour seals (Scholander *et al.*, 1942*a*), up to 20 min in Weddell seals (Kooyman *et al.*, 1980). Further diving, or diving with vigorous exercise, would depend upon anaerobic energy sources.

Diving bradycardia

The bradycardia of diving mammals has been observed in a wide range of other species including muskrat (Koppanyi & Dooley, 1929), sea lion (Elsner *et al.*, 1964*a*), fur seal (Irving *et al.*, 1963), manatee (Scholander & Irving, 1941), dugong (R. Elsner & D. D. Hammond, unpublished observations), porpoise (Irving, Scholander & Grinnell, 1941*a*), platypus

16

(Johansen, Lenfant & Grigg, 1966), water vole (Clausen, 1964) and water shrew (Calder, 1969). It has also been observed during free dives of grey seals (Scholander, 1940), Weddell seals (Kooyman & Campbell, 1972) harbour seals (Jones *et al.*, 1973), a killer whale (Spencer, Gornall & Poulter, 1967), pilot whale and grey whale (R. Elsner & D. W. Kenney, unpublished observations) and a hippopotamus (Elsner, 1966). It was recognized early that the responses to diving could be induced by head immersion alone and that apnoea combined with face wetting were the necessary conditions for their demonstration. The heart slowing in the diving animals is demonstrable early in life, but diving ability is not well developed in new-born and infant animals (Harrison & Tomlinson, 1960; Irving *et al.*, 1963; Kooyman, 1968; Hammond *et al.*, 1969).

Elsner *et al.* (1966*b*) demonstrated profound bradycardia in two Pacific bottle-nosed dolphins *Tursiops truncatus gilli* trained to dive on command to the bottom of a 2-m pool. In this way dives up to 4 min 42 s in duration were obtained. The heart rate fell promptly from about 90 beats/min to 12 beats/min, and slowly increased later to approximately 20 beats/min as the dive continued. Fig. 2.2 illustrates heart rates determined electrocardiographically in one of these free-swimming and diving dolphins. Oscillations of the heart rate with the respiratory cycles, increasing with inspiration, are clearly apparent. Similar marked sinus arrhythmia has been noted in other marine mammals (Irving *et al.*, 1935; Scholander, 1940; Bartholomew, 1954).

In laboratory studies on diving species, some means of restraint by which the animal can be forcibly submerged have generally been used. With careful training, seals and ducks can usually become accustomed to the procedure and remain relaxed during dives of moderate duration. Porpoises and dolphins, however, do not easily adapt to the procedure, struggle violently, and sometimes have even died during an experiment (Scholander, 1940; Irving *et al.*, 1941*a*).

Harbour seals *Phoca vitulina* and sea lions *Zalophus californianus* have been trained to dive by head immersion in a container of water (Elsner, 1965). The harbour seals were also trained to dive freely in a pool upon command. By this means

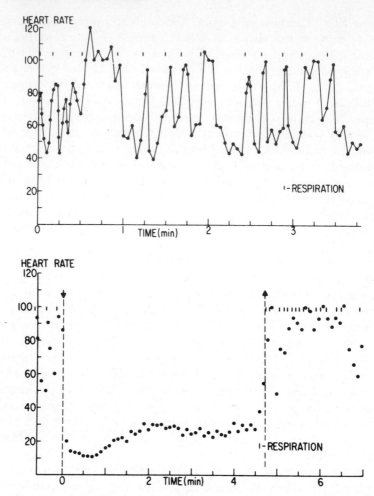

Fig. 2.2. Heart rate (beats/min) of a dolphin *Tursiops truncatus* during free swimming (above) and during a trained dive (below). Beginning and end of dive indicated by arrows. (Elsner *et al.*, 1966*b*.)

P. vitulina could be observed in trained dives of up to $7\frac{1}{2}$ min duration. *Z. californianus* did not dive for as long, the maximum duration of a trained dive being 90 s. The onset of bradycardia was slower in the trained dives than during forced dives of the same animal. The lowest value reached in *P. vitulina* was commonly 18–25 beats/min compared with

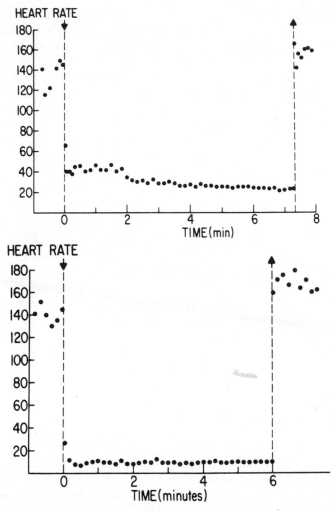

Fig. 2.3. Heart rate (beats/min) of a harbour seal *Phoca vitulina* during a trained dive (above) and during a restrained dive (below). Beginning and end of dives indicated by arrows. (Elsner, 1965.)

6–8 beats/min during forced immersion. Typical results are presented in Fig. 2.3. Ridgway, Carder and Clark (1975) obtained similar results with trained sea lions. They also succeeded in training a sea lion to lower its heart rate to 10 beats/min without diving.

The electrocardiogram of the diving animal shows some progressive changes during prolonged apnoeic dives. These include, besides the general bradycardia with prolongation of diastole, sometimes a gradual diminution of the P wave and occasionally its complete elimination. Cardiac rhythm is then apparently set independently of the sino-atrial node and is established by a ventricular pacemaker site. Cardiac arrhythmias occasionally appear on the electrocardiogram.

Bradycardia during diving, breath-holding and asphyxia has been observed in many vertebrate species, including man. Among those studied are rabbit (Bauer, 1938), sloth (Irving, Scholander & Grinnell, 1942b), snake (Johansen, 1959; Murdaugh & Jackson, 1962), dog (Elsner *et al.*, 1966a), pig (Irving, Peyton & Monson, 1956), armadillo (Scholander, Irving & Grinnell, 1943), echidna (Augee *et al.*, 1971) and lizards (Wood & Johansen, 1974). Terrestrial mammals generally do not respond as immediately nor as markedly as the aquatic species.

The significance of the diving bradycardia remained obscure long after its original observation. It seemed clear, however, that a heart slowing of such magnitude must be accompanied by a decrease in cardiac output, since an increase in stroke volume could hardly compensate for the low rate.

Cardiac output

Determination of cardiac output in the diving harbour seal by use of a dye dilution technique indicated that it declined in proportion to heart rate (Wasserman & MacKenzie, 1957; Murdaugh *et al.*, 1966). Elsner *et al.* (1964a) measured cardiac output by directly recording blood flow in the pulmonary artery of a sea lion which was trained to immerse its head in water. A chronically implanted ultrasonic flowmeter provided instantaneous measurement of blood flow. Heart slowing occurred, and the stroke volume remained unchanged during immersion. Similar determinations were made on dogs to test the response to immersion in terrestrial animals (Elsner *et al.*, 1966a). Blood flow was measured in the ascending thoracic aorta. As in the sea lion, stroke volume remained unchanged throughout the dive. High resting values of cardiac output in harbour seals were interpreted by Sinnett, Kooyman & Wahrenbrock (1978) as being consistent with their high metabolic rates. Using a thermal dilution technique, they also showed a one-third reduction in stroke volume during forced dives. Pulmonary blood flow may cease during diastole in such dives (Sinnett, Kooyman & Wahrenbrock, 1978).

Techniques using radioactive microspheres for the determination of cardiac output have shown an approximately 90% reduction in experimentally dived Weddell seals (Zapol *et al.*, 1979) and spotted seals (Elsner, Blix & Kjekshus, 1978; Kjekshus *et al.*, 1982) and about a 30% reduction in stroke volume. Microsphere trapping indicated that a considerable arteriovenous shunting of cardiac output occurred early in the dive. Later, however, cardiac output was routed through systemic capillaries mainly in the cerebral circulation, and a small fraction passed through peripheral A–V shunts (Blix, Elsner & Kjekshus, 1983).

Arterial pressure

The maintenance of arterial blood pressure is essential if adequate perfusion of vital organs is to take place during a profound decrease in heart rate and cardiac output. Irving, Scholander & Grinnell (1942*a*) measured the pressure in the femoral artery of *P. vitulina* during experimental dives. The mean arterial pressure remained almost constant despite the prompt heart slowing from 110 beats/min to 9 beats/min (Fig. 2.4, top). Elsner *et al.* (1966*a*) obtained measurements of blood pressure in the abdominal aorta of an elephant seal *Mirounga angustirostris* during restrained diving. One such record, obtained on a young adult male weighing 320 kg, is shown in Fig. 2.4, bottom. The intense bradycardia, sometimes as low as 6 beats/min, can be seen. Systolic pressure, which before the dive was about 110–130 mmHg, increased during the dive and varied from about 120 mmHg to 160 mmHg when the animal remained quiet. Reacting to peripheral vasoconstriction, pressure fell very slowly during the much-prolonged diastolic intervals. It was maintained at

21

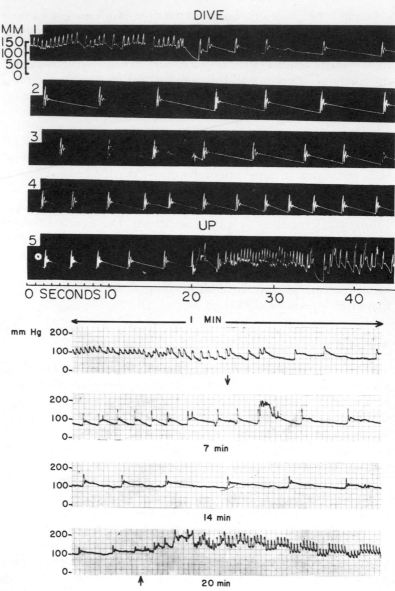

Fig. 2.4. Arterial blood pressure of, above, a restrained diving harbour seal *Phoca vitulina* (from Irving *et al.*, 1942*a*) and, below, an elephant seal *Mirounga angustirostris* (Elsner *et al.*, 1966*a*). Beginning and end of dives indicated.

22

about 80–100 mmHg, a value similar to that of the resting, non-submerged state. Thus mean arterial pressure remained almost unchanged despite the slow heart rate.

The energy for maintaining arterial pressure is provided by the stretching of the elastic arterial walls during and immediately following systole and its recoil during diastole. This 'Windkessel' function is augmented in many species of marine mammals by a bulbous enlargement of the root of the aorta, the 'aortic bulb' (Elsner *et al,*, 1966*a*). The aortic bulb approximately doubles the diameter of the ascending aorta in harbour and Weddell seals, providing a compensating reservoir for maintaining pressure and flow into the constricted arterial tree during the long diastolic intervals characteristic of diving (Drabek, 1975, 1977; Rhode, Peterson & Elsner, 1977). Terrestrial mammals which lack a structure comparable to the aortic bulb have relatively little volume and energy storage capacity at that site. Accordingly, the entire human aorta contains less volume than the aortic bulb alone in seals of similar body weight (Rhode *et al.*, 1983). Theoretical analysis of the properties of the seal's aortic bulb suggests another adaptive function. Its distensibility characteristics suggest that the work of the left ventricle against downstream impedance is about one-half to one-third less than it would be without such a structure. Thus, the increase in left ventricular afterload which would be expected as a consequence of elevated peripheral resistance and decreased large artery compliance is reduced. The result diminishes peak systolic pressure development, thus tending to lessen cardiac work and oxygen consumption while at the same time maintaining stroke volume (Campbell *et al.*, 1981; Rhode *et al.*, 1983).

Blood flow

The elucidation of dynamic and quantitative blood flow changes awaited the development of modern electronic blood flow measuring equipment which could be chronically implanted in experimental animals. Ultrasonic flowtransducers can be surgically implanted around an individual artery without puncture of the vessel (Franklin, Schlegel & Watson, 1963). Within a few days the animal is sufficiently

23

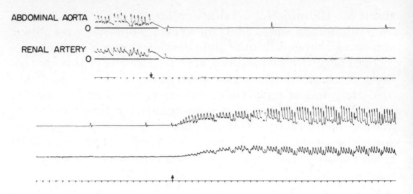

Fig. 2.5. Relative blood flow in the abdominal aorta and renal artery of a restrained diving harbour seal *Phoca vitulina*. Arrows indicate beginning and end of a 7-min dive. Time marks in s. (Elsner *et al.*, 1966*a*.)

recovered to permit experimental manipulation. By such an approach blood flow distribution during experimental dives in harbour seals and laboratory dogs was measured and compared by using recordings from the renal artery, terminal abdominal aorta and common carotid artery in harbour seals and the superior mesenteric artery, renal artery, and terminal abdominal aorta in dogs (Elsner *et al.*, 1966*a*; Elsner, 1969).

Fig. 2.5 illustrates the result of one experiment. In this instance the seal was experimentally immersed for 7 min. The animal was lightly restrained and had become accustomed to the procedure, lying quietly throughout the dive. The bradycardia and simultaneous abrupt fall of blood flow almost to zero at the beginning of the dive are evident. Similar indications of blood flow redistribution were recorded from chronically implanted transducers in diving penguins (Millard, Johansen & Milsom, 1973).

It was mentioned earlier that the response of the heart rate to diving was more profound during forced submersion than during a dive performed by an unrestrained animal. Similarly, when a seal was trained to dive on command for a brief interval, the blood flow decrease was less intense. However, blood flow in the abdominal aorta was still considerably diminished by the reduction of steady diastolic flow (Fig. 2.6) (Elsner *et al.*, 1966*a*).

24

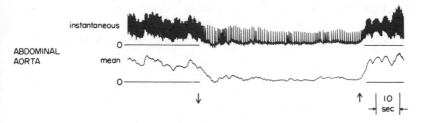

Fig. 2.6. Relative blood flow in the abdominal aorta of a harbour seal *Phoca vitulina* trained to dive on command. Arrows indicate beginning and end of dive. (Elsner *et al.*, 1966*a*.)

The blood flow distribution characteristics of a diving seal are not limited to aquatic animals as is shown by qualitatively similar but less intense responses in dogs. Elsner *et al.* (1966*a*) immersed the nose and mouth of previously instrumented dogs in water for 1 min. Blood flow declined slowly, had virtually ceased within 30 s in the superior mesenteric and renal arteries and was reduced to about 5% of the pre-dive level in the abdominal aorta. Flow remained at that level despite occasional struggling by the dog. Arterial pressure increased by about 50% during the episode. Flows returned promptly to their pre-dive values during recovery.

In other experiments instrumented dogs were treated with atropine that was intended to block the parasympathetic vagal innervation of the heart; 0.1 mg/kg was administered and resulted in an increase of heart rate from 150 beats/min to about 200 beats/min. This tachycardia persisted unchanged during brief snout immersion. However, despite the tachycardia, the blood flow in the mesenteric and renal arteries and the abdominal aorta was still much reduced, although with some delay. The results suggest that the administration of atropine produced a separation of the sympathetic and parasympathetic responses, sympathetic vasoconstriction still being elicited while parasympathetic bradycardia was abolished. A similar dissociation of these two components of the diving response was demonstrated during diving in harbour seals by intracardiac pacing, which resulted in vasoconstriction in the absence of bradycardia (Murdaugh *et al.*, 1968).

25

Distribution of cardiac output in a manner similar to that in aquatic mammals and birds was reported in diving rats (Lin & Baker, 1975) and asphyxiated rabbits (Wyler & Hof, 1977).

The echidna *Tachyglossus aculeatus*, an Australian monotreme, burrows into the ground when frightened or when seeking ants and termites for food. Laboratory studies in which echidnas burrowed into a large metal bin filled with soft earth showed that the animal develops bradycardia when exposed to this form of partial asphyxia (Augee *et al.*, 1971). The bradycardia was elicited by increasing carbon dioxide concentrations in its inspired air and was less responsive to low oxygen concentrations. Echidnas can swim, although they are not likely to encounter much water in their arid habitat. When an echidna was experimentally submerged, its heart rate and abdominal aorta blood flow were much reduced. Similar responses were seen during the animal's decline of body temperature when in torpor.

Since the argument for oxygen sparing depends upon there being continued blood flow in heart and brain, it is useful to consider the evidence for such a condition during asphyxia. Existing information bolsters the argument, but most measurements are indirect and do not permit rigorous and dynamic examination of processes and controls. Using the fractional distribution of circulating radioactive rubidium, Johansen (1964) determined the partition of cardiac output in the diving duck. The heart, head, eye and adrenal glands all showed relatively high activity.

In a study of cerebral tolerance to asphyxial hypoxia, Kerem & Elsner (1973*b*) showed that blood flow in the extradural intravertebral vein, which drains blood from the brain, in the dived harbour seal remained at about 75% of its pre-dive value, increasing slightly late in the dive. The flowmeter, a catheter-tip ultrasonic device, was positioned immediately below the exit of venous drainage from the skull. Although the major portion of cerebral venous drainage probably flows through that large conduit, an undetermined fraction passes through the internal jugular system, despite its relatively small size in this species. In contrast with the harbour seal, cerebral blood flow in sea lions increased steadily during diving (Dormer, Denn & Stone, 1977).

Determinations of blood flow distribution by circulatory

trapping of radioactive microspheres has provided useful information about the relative perfusion of organs in diving seals. Blix *et al.* (1976a) showed that myocardial blood flow in the dived grey seal *Halichoerus gryphus* decreased to 10% of the blood flow in the pre-dive condition. Cerebral flow was at the same time reduced to 34% of its former value. Flow was almost completely abolished in skeletal muscle, pancreas, liver, spleen, thyroid and kidney of Weddell seals during restrained dives (Zapol *et al.*, 1979). Similarly treated spotted seals showed a decline of about 50% in cerebral cortical flow after 5 min of diving, but after 10 min it had increased above the pre-dive value. Brain stem perfusion remained less than the control level. Liver, kidney, spleen, pancreas, stomach, intestine and subcutaneous fat had little or no detectable flow. Skeletal muscle was perfused during the resting, pre-dive condition at a low rate (3 ml/100 g min) and had virtually no flow during diving. Muscle perfusion during recovery was highly variable (5–105 ml/100 g min) at different locations (Elsner, Blix & Kjekshus, 1978; Blix, Elsner & Kjekshus, 1983). The determinations of flow in skeletal muscle made during recovery from the dive indicate that restoration of normal flow did not occur uniformly throughout that tissue. This response is likely to play an important role in the maintenance of systemic arterial pressure, which would otherwise be threatened by the low resistance of fully dilated arterial channels in skeletal muscle.

In addition to heart and brain, the adrenal gland also continues to receive considerable blood flow during diving, although perfusion appears to be somewhat below normal levels (Johansen, 1964; Elsner, Blix & Kjekshus, 1978; Jones *et al.*, 1979; Zapol *et al.*, 1979). Liggins *et al.* (1979) reported high levels of adrenal corticosteroid in the circulation of Weddell seals. Uterine blood flow in the pregnant Weddell seal is maintained during diving (Elsner, Hammond & Parker, 1970a; Liggins *et al.*, 1980).

Venous structure and function

Early workers who described the anatomy of diving animals were impressed by the unusual blood storage capacity of the venous system. The historical development of anatomical

Metabolic conservation

knowledge of marine mammals has been interestingly reviewed by Harrison (1972). Hunter (1787) described the large blood volume in whales and seals and mentioned the dense thoracic arteriovenous networks in whales. These retia have a pressure-damping effect on cerebral circulation (Nagel *et al.*, 1968; Viamonte *et al.*, 1968) but whatever adaptive function they may have remains a mystery. Burow (1838) provided a detailed description of the large-capacity venous system in seals. Murie (1874) remarked on the great capacity and complication of the venous system in the sea lion. Gratiolet (1860) described the enlarged inferior vena caval system of the hippopotamus and its sphincter at the level of the diaphragm. More recently, Harrison and Tomlinson (1956) reviewed present knowledge of the anatomy of the venous system in pinnipedia and cetacea. Blessing & Hartschen (1969) described the caval sphincter in harbour seals and Ronald, McCarter & Selley (1977) discussed the venous morphology of harp seals. The capacity of the venous structures of several species of marine mammals can provide for unusually large storage of blood.

The venous system of seals differs from that of other mammals by several important features: (1) the inferior vena cava is commonly bifurcated and can hold a relatively large volume of blood, (2) the inferior vena cava enlarges to form hepatic sinuses of unusual size, (3) venous drainage from the kidney is accomplished by means of diffuse venous sinuses rather than a discrete renal vein, (4) cerebral venous drainage is accomplished mainly by a large vein lying dorsal to the spinal cord and connecting with renal veins and the inferior vena cava and (5) a sphincter composed of striated muscle is located in the inferior vena cava at the diaphragm in several species.

The inferior vena cava of adult elephant seals *M. angustirostris* is of such volume as to contain approximately one-fifth of the total blood volume (Elsner *et al.*, 1964b). The data of Fig. 2.7 illustrate the relation between arterial and venous blood gases during diving. Sampling from the abdominal aorta and the inferior vena cava showed that the depletion of blood oxygen in the arterial circulation was accompanied by a slower decline of oxygen in the venous reservoir. Sometimes

28

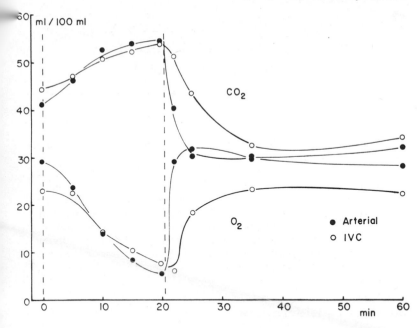

Fig. 2.7. Oxygen and carbon dioxide contents of aortic and inferior vena caval (IVC) blood of a diving elephant seal *Mirounga angustirostris*. Restrained dive occurred between vertical dashed lines. (From Elsner, 1969.)

in the latter portion of the dive blood oxygen was higher in the inferior vena cava than in arterial circulating blood. The blood levels of carbon dioxide mirrored this phenomenon. Harrison & Tomlinson (1956) demonstrated that the inferior vena cava sphincter in *P. vitulina* is innervated by a branch of the right phrenic nerve and constricts when stimulated. Radiographic evidence indicates that venous blood flow returning to the heart by this route is restricted during dives by action of the muscular sphincter (Elsner, Hanafee & Hammond, 1971*c*; Hol, Blix & Myhre, 1975). Such a mechanism would help to limit cardiac output and adjust its magnitude to the reduced diving requirement. Right atrial pressure remained unchanged in restrained dives of harbour seals (Sinnett *et al.*, 1978).

29

3 CELLULAR TOLERANCES AND ADAPTATIONS TO ASPHYXIA

About the middle of the last century the younger physiologists broke away from the vitalistic traditions which had been handed down to them, and set about to investigate living organisms piece by piece, precisely as they would investigate the working of a complex mechanism . . . It is still the orthodox method of physiology, but the old confidence in it has steadily diminished in proportion as exact experimental investigation has shown that the various activities of a living organism cannot be interpreted in isolation from one another, since organic regulation dominates them (J. S. Haldane, 1922).

Survival considerations

Acute but transient loss of function in many structures does not present an irreversible threat to the survival of the animal. Man can withstand temporary loss of function of the kidneys, liver or gut, for example, without placing the survival of the whole body in jeopardy. Similarly, sudden loss of function of one or more limbs is not ordinarily fatal. On the other hand, a sudden, unexpected loss of central nervous function can lead rapidly to an irreversible situation ending in death. Cessation of the normal pumping activity of the heart is rapidly followed by ischaemia and asphyxia of the cerebral tissue with loss of consciousness. Clearly, the heart and brain are the most important tissues for the short-term survival of the organism when faced with the threat of acute asphyxia. Other body tissues are functionally impaired to varying degrees by asphyxia, but they are relatively more resistant to permanent damage. These considerations suggest that when an animal is faced with an acute asphyxial threat, it would be an adaptive advantage, both in the short and the long term, to protect the heart and brain against asphyxia at the expense of other tissues. This basic concept was outlined by Franklin (1951) in

30

the Oliver-Sharpey Lectures under the title 'Aspects of the Circulation's Economy'.

Resistance of organs and tissues to asphyxia

When the blood flow to an organ or tissue is arrested, its oxygen tension falls, its carbon dioxide tension rises above normal levels and acidity increases. The cells thus exposed to progressive asphyxia continue to perform their specific functions for a variable period depending upon their capacity for anaerobic metabolism, but eventually this functional limit is reached. However, with depression of normal activity the cells can survive for a longer period, especially at lower than normal temperatures. If during this time blood flow is restored, the tissue can recover its functional capacity without permanent impairment. The resistance of different cells to asphyxia varies widely (Table 3.1). Some organs of habitual divers have unusual tolerance to ischaemia and asphyxia. Comparisons with terrestrial animals, including humans, can be instructive.

Oxygen balance sheet in marine mammals

Several calculations have been made of oxygen storage, which is dependent upon blood volume and haemoglobin concentration, and its estimated utilization in diving seals. The values are only approximations, and few conclusions can be reached regarding precise utilization and requirements during a dive (Scholander, 1940; Packer *et al.*, 1969; Lenfant, Johansen & Torrance, 1970; Kooyman, 1975). General agreement is that non-lung oxygen stores for the harbour seal amount to approximately 30 ml/kg. Comparisons of blood oxygen storage in harbour seal, Weddell seal and man are shown in Fig. 3.1. Brain weight in the harbour seal represents approximately 0.4% of body weight, and the heart weight is about 0.7% of body weight. Since these organs constitute the major portions of the metabolizing and perfused body during diving, considerations of oxygen availability and possible adaptations for declining oxygen use will be referred to them. A general trend is discernible indicating that marine mammals within the same family show a gradation in maximum

Table 3.1 *Resistance of mammalian tissues to warm ischaemia.*

Tissue	Survival (in min)	References
Brain	5–10 (man) (cortex)	Fazekas & Bessman (1953) (clinical observations)
	6–8 (dog)	Kabat, Dennis & Baker (1941); Grenell (1946)
Heart (resting)	4.5 (dog)	Schlosser & Streicher (1964)
Liver	20 (dog)	Raffucci & Wagensteen (1951)
	30–45 (rat)	Baker (1956)
Kidney	60 (dog)	Birkeland *et al.* (1959)
	75 (dog)	Cohn & Moses (1966)
Skeletal muscle (resting) and skin	780–900 (rat, rabbit, cat, dog)	Allen (1938*a*)

diving time, the larger members being capable of longer durations than the smaller. Their brain weights represent a smaller fraction of total body weight, and this characteristic might account for the differences (Ferren & Elsner, 1979).

Aerobic oxidative metabolism in skeletal and heart muscle in some marine mammals is supported by high concentrations of myoglobin. In phocid seals myoglobin represents an oxygen storage of roughly five times the quantity available to terrestrial species (Scholander, 1940; Scholander *et al.*, 1942*a*). The storage thus achieved is considerable, since skeletal muscle constitutes about 40% of the fat-free body mass. Muscle perfusion is generally much reduced or entirely eliminated during long or forced dives, and consequently the utilization of myoglobin-bound oxygen is restricted to skeletal muscle. Distribution of myoglobin among species is variable and depends upon many factors (Kagen, 1973). Its content is less in the muscle of new-born animals than in that of adult animals (Kagen & Christian, 1966) and is usually increased by chronic exposure to the hypoxia of high altitudes and by exercise (reviewed by Kagen, 1973). Among marine mammals, moreover, considerable variation exists; common porpoise *Phocaena phocaena* muscle content is about ten times

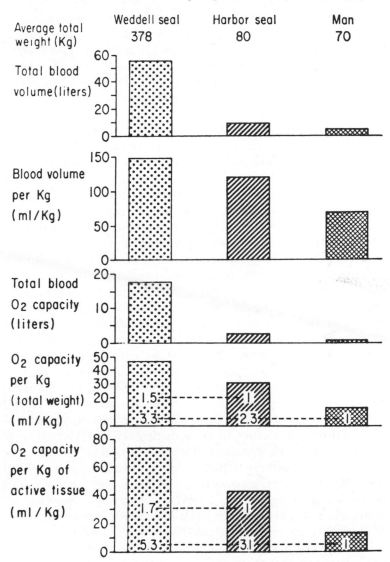

Fig. 3.1. Blood oxygen storage in Weddell seal *Leptonychotes weddelli*, harbour seal *Phoca vitulina* and man. Relative blood oxygen capacity of seals compared to man is indicated. (From Lenfant *et al.*, 1969*b*.)

higher than in the manatee *Trichechus manatus* (Blessing, 1972). Seals, porpoises and manatee all have higher cardiac myoglobin levels than do humans (Blessing & Hartschen-Niemeyer, 1969; Blessing, 1972). Reference was made earlier (Chapter 1) to the dependence of skeletal muscle in many marine mammals on oxygen derived from myoglobin during aerobic dives.

Selected summary data on oxygen storage in marine mammals have been compiled in Table 3.2. Generally speaking, phocid seals, whose dives are the longest among the pinnipedia, have high values for blood and muscle oxygen stores, while the sea otter, a relatively shallow and brief diver, relies chiefly on lung storage. The breathing pattern of marine mammals is sometimes slow and irregular, producing fluctuations in blood gases in a grey whale (Wahrenbrock *et al.*, 1974) and harbour seals (Sinnett *et al.*, 1978). No consistent adaptive pattern appears in the respiratory properties of blood, for example in the affinity of haemoglobin for oxygen and the Bohr effect. However, the blood of some natural divers has an enhanced Bohr effect which may contribute to continued oxygen extraction through the mechanism of lowering the affinity of haemoglobin for oxygen as carbon dioxide tension increases. Buffering capacity of blood is elevated in marine animals (Fig. 3.9) (Lenfant *et al.*, 1970; Wood & Johansen, 1974). Myoglobin probably plays an important role in the buffering of acid metabolites formed in the inadequately perfused muscle tissues of seals during diving (Castellini & Somero, 1981). Restoration of normal blood oxygenation is aided by a relatively small lung residual volume which provides for more complete pulmonary gas exchange in porpoises (Irving *et al.*, 1941*a*; Ridgway, Scronce & Kanwisher, 1969), seals (Scholander, 1940; Kooyman *et al.*, 1971, 1973) and pilot whales (Olsen *et al.*, 1969).

The elevated oxygen storage capacity of phocid seal blood might suggest that these species are preacclimatized to high altitude, an environment which they would not be expected to encounter. However, two harbour seals showed rather typical mammalian ventilatory and erythropoietic responses during a sojourn of 85 days at 3100 m. New-born harbour seals at sea level showed a steady increase in blood volume and circulat-

Table 3.2. Blood and muscle properties related to oxygen storage

Species	Haematocrit (%)	Haemoglobin (g/100 ml blood)	Blood volume (ml/kg)	Myoglobin concentration (g/100 g muscle)	Reference
Harbour seal	58	21	132	5.5	Lenfant, Johansen & Torrance (1970)
Ribbon seal	67	24	132	8.1	Lenfant, Johansen & Torrance (1970)
Fur seal	49	17	109	3.5	Lenfant, Johansen & Torrance (1970)
Sea lion	45	15	92	2.4	Lenfant, Johansen & Torrance (1970)
Walrus	42	16	106	3.0	Lenfant, Johansen & Torrance (1970)
Sea otter	48	17	91	2.6	Lenfant, Johansen & Torrance (1970)
Beaver	–	12	6.5	1.2	McKean & Carlton (1977)
Water vole	46	14	–	–	Clausen & Ersland (1968)
Ringed seal	55	24	234[a]	–	Ferren & Elsner (1979)
Weddell seal	61	24	141	–	Lenfant et al. (1969a)
Domestic hen	34	11	–	–	Christensen & Dill (1935)
Domestic duck	–	16	–	–	Christensen & Dill (1935)
Adélie penguin	46	17	93	–	Lenfant et al. (1969c)
Emperor penguin	47	17	–	–	Lenfant et al. (1969c)
California murre	50	15	–	–	Lenfant et al. (1969c)

[a] Expressed as ml/kg lean body weight.

ing haemoglobin during the first year of growth. Seals which were prevented from diving throughout the first year of life achieved during that time blood oxygen capacities which were comparable to normal seals (Kodama, Elsner & Pace, 1977). Furthermore, the naive seals responded to their first dives at one year of age with physiological reactions that were undistinguishable from those of experienced seals (Elsner, 1979). Thus a phylogenetic rather than an environmental basis for maturation of diving responses is indicated.

Brain

The vertebrate brain is notably sensitive to oxygen deficiency. While the evolutionary development of air breathing and stable body temperature have assured mammals and birds a certain freedom from environmental constraints, the threats of pathological interference with respiration are only a breath away. There are three exceptions to the general mammalian intolerance to cerebral hypoxia: fetal and newborn animals (Dawes, Mott & Shelly, 1959), hibernators (Bullard, David & Nichols, 1960) and at least some diving mammals, for example Weddell seals (Elsner *et al.*, 1970*b*) and harbour seals (Kerem & Elsner, 1973*b*).

The first suggestion that the brain of seals may have special tolerance to low oxygen concentrations was given by the observation of Scholander (1940) that the extreme minimum of arterial blood oxygen in a long dive by grey and hooded seals reached a value of 2 ml/100 ml (Fig. 3.2). Conversion of that content to partial pressure suggests that it would amount to approximately 10 mmHg, allowing for some assumptions regarding the pH of the animal's blood. Simultaneous measurement of arterial oxygen partial pressures and electro-encephalographic (EEG) recordings, noting the appearance of hypoxic EEG high-amplitude, slow-wave activity of Weddell seals (Elsner *et al.*, 1970*b*) and harbour seals (Kerem & Elsner, 1973*b*) during diving, showed that these species were able to tolerate a decline in arterial blood oxygen to levels (8–10 mmHg) which would not be compatible with normal brain function in terrestrial mammals (Meyer *et al.*, 1962). The EEG endpoint remained unchanged at the low oxygen tension despite a wide range of blood P_{CO_2} and pH

36

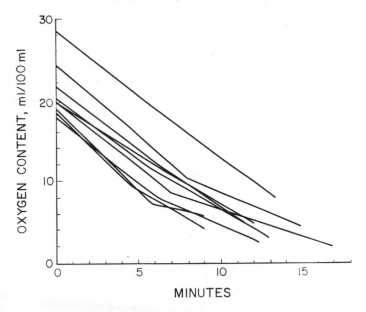

Fig. 3.2. Oxygen depletion from arterial blood of diving grey seals *Halichoerus gryphus* and hooded seals *Cystophora cristata*. (Redrawn from Scholander, 1940.)

values. The capillary density in seal brain tissue is about 50% higher than in cats and humans and suggests that a sufficiently high density of the capillaries through which nutritive blood is perfused could compensate for declining oxygen partial pressure by shortening the diffusion distances to cells. Late in the dive brain oxygen consumption declined and lactic acid appeared in the venous blood draining that organ, signalling a conversion to anaerobic glycolysis, an unusual condition in terrestrial mammalian brain tissue (Kerem & Elsner, 1973*b*).

The length of time that the human brain can be temporarily deprived of its circulation and subsequently recover completely is difficult to determine precisely. Rossen, Kabat & Anderson (1943) found that sudden arrest of the cerebral circulation produced loss of consciousness within 6 or 7 s. Restoration of flow within 100 s was followed by a rapid return of consciousness with no evidence of brain injury.

Complete but slow recovery of the brain has occurred after circulatory arrest lasting up to 6 min (Noble, 1946; Lampson, Schaeffer & Lincoln, 1948; Turner, 1950; Wolff, 1950). However, the outcome is highly variable and may depend upon whether or not cerebral atherosclerosis is present, Neubuerger (1954) reported on seven cases of arrest lasting from 1 to 10 min. All of the patients remained in a coma and died after an interval varying from 2 h to 2 weeks. According to Fazekas & Bessman (1953), the cerebral cortex becomes irreversibly damaged after 5–10 min of anoxia whereas the medulla may function effectively after anoxia of 20–30 min duration.

Grenell (1946) demonstrated a gradient of resistance to hypoxia in dog brain with the cerebral cortex having the lowest resistance, and the medulla oblongata having the highest. Arrest of the brain circulation for 6 min or less resulted in cerebral dysfunction which disappeared completely over a period lasting up to several days, but ischaemia for 8 min or longer resulted in permanent severe brain damage (Kabat, Dennis & Baker, 1941). Marshall, Owens & Swan (1956) found permanent histological damage in dogs' brains following 8 min of arrest. Functional recovery does not imply complete structural recovery as demonstrated by Grenell (1946), who found lingering severe brain damage in dogs that had made an apparent recovery from cerebral ischaemia.

Unanaesthetized, curarized dogs were subjected to periods of asphyxia by the arrest of artificial ventilation until the appearance of reversible hypoxic patterns on the electroencephalogram (Kerem & Elsner, 1973a). The average time to this endpoint was 4.5 min. The theoretical 'aerobic survival time' if total body metabolism remained at the pre-asphyxia resting level was calculated to be 4 min. At no time did a significant arteriovenous lactate gradient develop, suggesting that anaerobic glycolysis contributes little or nothing to dog brain metabolism during acute asphyxia.

Heart

The capability of seal heart muscle to continue functioning after arterial P_{O_2} has fallen to extremely low values leads one to suspect that that tissue may have specific tolerance to low oxygen concentrations and to local ischaemia. However, harbour seal heart was found to respond to local acute occlusions of coronary arteries in a manner similar to dogs (Elsner *et al.*, 1976). That is, acute regional ischaemia resulted in prompt failure of the ischaemic region to contract as vigorously as an adjacent non-ischaemic region, and this condition resulted in systolic bulging (Fig. 3.3). The technique used small crystals implanted in heart muscle in such a way that distances between them could be electronically determined and recorded (Theroux *et al.*, 1974). Post-mortem examinations of seal heart by injection of the coronary circulation showed that the density of coronary anastomoses, the anastomotic index (Menick, White & Bloor, 1971) is approximately one-tenth of the value found in dogs.

Coronary blood flow in mammals is regulated to serve the metabolic requirements of the myocardium, which are determined by developed wall tension (derived from intraventricular pressure and ventricular end-diastolic volume), myocardial contractility and heart rate (Braunwald, 1971). Hypoxaemia and anaerobiosis normally impose a profound stress on the regulation of myocardial blood flow and metabolism (Lekven, Mjøs & Kjekshus, 1973). Submersion of diving seals results in a reduction of arterial oxygen content and coronary flow (Blix *et al.*, 1976a; Kjekshus *et al.*, 1982). Thus the myocardium would become ischaemic if oxygen requirements remained unchanged. Increased anaerobic metabolism could, however, compensate for some of the diminished oxidative metabolism. In terrestrial animals this is usually of very limited value owing to its small contribution relative to the total energy requirement and the rapid inactivation of glycolytic enzymes in the ischaemic heart (Rovetto, Lamberton & Neely, 1975).

Despite the greatly reduced coronary blood flow during dives, there were no indications of myocardial ischaemia on electrocardiograms of harbour seals in long dives (Kjekshus *et al.*, 1982). Ventricular contractility (dP/dt_{max}) was reduced by 25–43% and ventricular end-diastolic pressure remained unchanged or increased only slightly late in the dive. Regional myocardial dimensions measured by implanted miniature

39

40

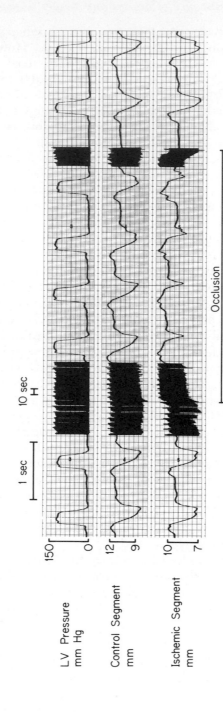

Fig. 3.3. Response of harbour seal *Phoca vitulina* heart muscle to acute local ischaemia indicating bulging of left ventricular wall. (Elsner, Franklin, White, McKown & Kemper, unpublished results).

crystal sonomicrometers showed variable and regular changes but no indication of sustained ventricular dilation (Millard *et al.*, 1980).

Perfusion within the mammalian heart muscle varies according to the transmural distribution of pressure during the cardiac cycle. Left ventricular wall blood flow resistance is highest in subendocardial regions and lowest in sub-epicardial regions (Kirk & Honig, 1964). Although intrinsic compensatory autoregulation generally brings about relative equality of perfusion of the inner and outer layers of the left ventricle wall, failure to achieve such distribution of flow is a consequence of increased end-diastolic pressure and related ventricular dilation resulting from ischaemic insult. Results of experiments on seals show that the ratio of subendocardial to subepicardial perfusion was unaltered, even during prolonged diving, indicating that the seal's heart maintained its functional integrity and was not ischaemic, despite its low blood flow, throughout the dives (Kjekshus *et al.*, 1982). Resting myocardial flow values were slightly higher than those observed in dogs by Kjekshus, (1973). Ventricular perfusion declined from an average of 150 ml/100 g min to 24 ml/100 g min. Electrocardiographic ST-segment elevations indicative of myocardial ischaemia were never observed, even in dives lasting up to 16 min. The average reduction of myocardial blood flow in the diving seal is of the same magnitude as the reduction of flow observed in the centre of a myocardial infarction caused by acute coronary occlusion in dogs. However, a reduction of blood flow in the dog similar to that observed in the seal results in extensive myocardial necrosis (Kjekshus, Maroko & Sobel, 1972).

Coronary blood flow was drastically reduced during forced diving and remained diminished but oscillating throughout the diving episode. Recent experiments using direct measurement of coronary blood flow by implanted flowmeters have shown more clearly what happens. Large, spontaneous oscillations in the left circumflex coronary artery flow occurred in the non-diving state, and experimental dives of a few minutes' duration resulted in intense vasoconstriction with flow nearly vanishing or totally eliminated for periods of 5–45 s. Flow was then restored at intervals of 15–45 s (Millard *et*

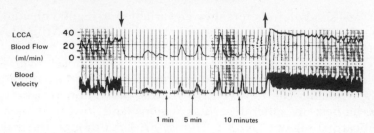

Fig. 3.4. Blood flow in the left circumflex coronary artery (LCCA) of a spotted seal *Phoca vitulina largha* during the first, fifth and tenth minutes of a dive (above). The instantaneous blood velocity record (ultrasonic Doppler flowmeter, below) shows deflections with heart beats. Beginning and end of dive indicated by arrows. (Elsner, Millard, Kjekshus & Blix, unpublished results.)

al., 1980; Elsner *et al.*, 1981*a*, *b*) (Fig. 3.4). The intense coronary vasoconstriction is reminiscent of a condition of vasospasm which occurs in some patients with coronary artery disease (Maseri *et al.*, 1978). The question of whatever similarities in governing mechanisms may exist in these two conditions remains unresolved. The response in seals somewhat resembles the competition between sympathetic vasoconstriction and metabolic vasodilation described by Mohrman & Feigl (1978) in dogs.

The coronary arteriovenous oxygen difference of seals declined steadily during dives (Kjekshus *et al.*, 1982), and the average oxygen consumption was reduced. It amounted to only 14% of pre-dive values early after submergence and was gradually lowered to 7% at the end of a 15-min dive. Myocardial lactate production occurred and increased as the dive continued. The reduction in myocardial oxygen consumption coincided with myocardial lactate release, suggesting a progressive shift from aerobic to anaerobic metabolism throughout the dive. A similar production of lactate caused by ischaemia in dogs is associated with coronary vasodilation as well as dilation of the heart (Scheuer & Brachfeld, 1966; Lekven, Mjøs & Kjekshus, 1973). Overall, the evidence shows that the seal heart is resistant to hypoxia if minimum blood flow is maintained, but it has poor tolerance to ischaemia. It adapts to diving asphyxia by a greatly reduced work load and consequent lowered oxygen requirement, and an enhanced capability for anaerobic metabolism. More detail is provided under *Biochemical Adaptations.*

The ability of canine hearts to tolerate complete ischaemia was examined by Schlosser & Streicher (1964). The critical duration of ischaemia at 37°C was 4.5 min. No animal survived ischaemia of 5 min duration for more than 12 h, and death was caused by cardiac failure. Tolerance to ischaemia was improved by lowering the temperature. Metabolic studies indicate that relatively short periods of ischaemia cause irreversible damage to the enzymes of aerobic energy production (reviewed by Katz, 1977). Respiration of heart muscle previously exposed to as little as 5 min of anoxia was reduced when oxygen was resupplied. Hearts from healthy people who died suddenly were revived up to 30 min after death (Kountz, 1936). However, the success rate fell sharply after that time interval. Only three out of eight hearts obtained 1 h after death could be revived.

Gooden, Stone & Young (1974) showed that mean coronary flow (measured by an ultrasonic flowmeter) in dogs trained to immerse their noses in water for 30 s was reduced by an average of 26%. Heart rate decreased by 48%. Myocardial oxygen consumption decreased by 42%. That decrease was abolished by cardiac pacing which prevented the immersion bradycardia. Ventricular contractility (dP/dt_{max}) was decreased late in the dive compared with pre-dive values. Ferrante & Opdyke (1969) have shown a decrease of 25–50% in left ventricular contractility in diving nutrias *Myocastor coypu,* and a similar negative inotropic effect has been observed in the heart of the diving duck (Folkow & Yonce, 1967). The possibility that vagal stimulation by immersion may produce a negative inotropic effect on the heart (decreased contractility, dP/dt_{max}) in diving animals is a matter of some controversy. Results of a study of ducks by Folkow & Yonce (1967) were so interpreted by them, but Furnival, Linden & Snow (1973) have challenged that conclusion. Their study suggests that duck ventricles may not be directly innervated by the vagus nerve. It would be premature to conclude that the decreased dP/dt_{max} seen in diving seals resulted from stimulation of the left ventricle through vagal innervation inasmuch as such neural connections have not been identified.

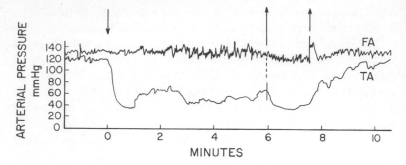

Fig. 3.5. Arterial blood pressure records from separate restrained dives, indicated by arrow, in a harbour seal *Phoca vitulina*. FA, femoral artery; TA, toe artery. (Redrawn from Irving *et al.*, 1942*a*.)

Arteries

One of the theoretical questions regarding mechanisms for maintenance of the diving response relates to how vaso-constriction in peripheral arteries can be maintained over long periods of maximum diving time. It is clear from studies on terrestrial species that metabolic vasodilator substances can successfully compete with neural vasoconstrictor in-fluences thus resulting in overriding of that regulation and vasodilation (Blair, Glover & Roddie, 1961; Kjellmer, 1965). The maintenance of vasoconstriction in diving seals suggests that those species may have specific vasoconstrictor reg-ulatory mechanisms somewhat modified from those of land mammals.

Irving *et al.* (1942*a*) recorded blood pressure in the femoral artery of a diving harbour seal and also in a small artery of the flipper having an inside diameter of 1 mm. While average femoral artery pressure changed little, the pressure in the small peripheral toe artery fell promptly and remained low until the end of the dive (Fig. 3.5). The steep pressure gradient from the femoral artery to a peripheral toe artery shows that the sites of vasoconstriction included arteries bigger than the small arterioles.

Radiographic evidence of constriction in vessels larger than arterioles can be seen in publications of Bron *et al.* (1966) and White, Ikeda & Elsner (1973). Morphological studies of

44

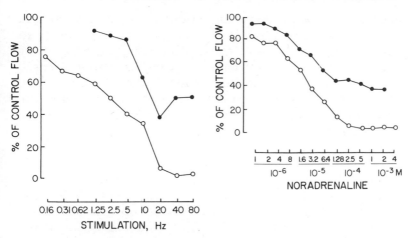

Fig. 3.6. Responses of chicken (filled circles) and duckling (open circles) mesenteric arteries to electrical stimulation and noradrenaline. (Redrawn from Gooden, 1980a.)

adrenergic nerve terminals on small arteries showed that the innervation in both ducks (Folkow, Fuxe & Sonnenschein, 1966) and seals (White *et al.*, 1973) extends to larger vessels in contrast to terrestrial species where the nerve terminals are concentrated on the arteriolar structures. Furthermore, nerve terminals penetrated into the media of the arterial wall in seals (White *et al.*, 1973). The presence of well-developed innervation in large arteries appeared to extend the range of vascular control from a metabolic competitive level at the arterioles in the natural divers.

   A detailed and quantitative comparison has been made of the structure and function of the isolated mesenteric arterial vasculature from the chicken (*Gallus domesticus*) and duckling (*Anas platyrhynchos*) (Gooden, 1980a). Vasoconstrictor responses to peri-arterial nerve stimulation and intra-arterial noradrenaline were evoked in tissues perfused at either constant flow rate or constant perfusion pressure. The duckling vasculature reacted with significantly greater responses to both nervous stimulation and noradrenaline (Fig. 3.6). Quantitative histological studies of the mesenteric arteries from the two species revealed two differences. First,

45

the mean wall thickness : lumen diameter ratios of the primary and secondary branches of the duckling mesentery were significantly greater than those of the chicken. Second, the main artery and its primary branches were more densely innervated in the duckling than in the chicken. These structural differences explain, at least partially, the differences observed in the vasoconstrictor power of the duckling arteries.

Kidney

Bradley & Bing (1942) were the first to show that diving resulted in profound reductions in the renal blood flow, glomerular filtration and urine production of seals. These results were confirmed by Bradley, Mudge & Blake (1954), Lowrence *et al.* (1956) and Murdaugh *et al.* (1961*a*), who found that those kidney functions ceased completely during 10-min dives.

The diving harbour seal has very low renal blood flow during dives, although the absolute level of flow is variable, depending upon the diving condition. It is clear from records obtained from implanted blood flowmeters that renal flow in a restrained experimental dive is virtually eliminated (Elsner *et al.*, 1966*a*). The question arises as to what the tolerance of ischaemia in seal kidneys might be compared with that of similarly treated dog kidneys. Isolated dog and harbour seal kidneys were perfused with oxygenated donor blood at 32 to 34°C. Perfusions were then stopped for 1 h. Blood flow in seal kidneys was restored nearly to pre-ischaemic values 15 min after perfusion was resumed, but dog kidney flow was drastically reduced after ischaemia and remained at or below 50% of the control level. All seal kidneys promptly recovered urine production while all dog kidneys remained anuric. Seal renal function (creatinine filtration fraction and *p*-aminohippuric acid (PAH) extraction ratio) was only slightly decreased after ischaemia (Halasz *et al.*, 1974).

Other evidence of tolerance to low oxygen tensions in seal kidneys was obtained by studies of organic anion transport in renal cortical slices of harbour seal kidneys. Transport of PAH was not altered after 60 min of incubation in nitrogen (Koschier *et al.*, 1978). Furthermore, renal organic ion transport and sodium/potassium exchange pump activities in seal kidney slices were more resistant to anoxia than in rat kidney, and seal renal organic anion transport was also tolerant of low pH (Hong *et al.*, 1981, 1982).

Dutz & Kretzschmar (1954) reported no urine flow in dogs for 1 h after 45 min of ischaemia. Clearances of inulin, PAH and urea were all less than 50% of control levels 90 min after ischaemia. Phillips & Hamilton (1948) described the results of 60 min of warm ischaemia studied 90 min after reflow as diminution of the PAH extraction ratio by 66% and of the creatinine filtration fraction by 72%. Friedman, Johnson & Friedman (1954) studied patterns of recovery from 2 h of ischaemia 3 or 4 h later and found creatinine clearance to be 3.3% and PAH clearance 2.8% of control levels.

Liver

The limit of tolerance to ischaemia of rat liver cells at body temperature appears to be 30–45 min (Baker, 1956). The animals were sacrificed 18 h after selected liver lobes had been rendered ischaemic, and tissue from ischaemic lobes showed histological evidence of necrosis. However, in dogs Raffucci & Wangensteen (1951) found that 20 min was the maximum period of occlusion of hepatic artery and portal vein tolerated at normal body temperature without indication of damage. In the hypothermic dog longer periods were tolerated with minimum functional or structural injury (Raffucci, Lewis & Wangensteen, 1953).

Liver blood flow is markedly reduced in restrained seals during dives (Elsner *et al.*, 1978; Zapol *et al.*, 1979).

Limbs

The upper limit of resistance of human limbs to asphyxia is unknown (Editorial, Can. Med. Assoc. J., 1973). The maximum recommended time for application of a tourniquet varies from 1 to 3 h and appears to be largely empirical (Bruner, 1970; Parkes, 1973). Flatt (1972) reviewed 1500 hand operations in which a tourniquet was applied. In 60 patients aged from one to 64 years, the tourniquet time exceeded 2 h, but none of these patients showed any post-operative complications. Shaw Wilgis (1971) obtained sam-

47

ples of blood at intervals from human limbs being operated on under a tourniquet over 2 h. After 2 h had elapsed, the P_{O_2} was 0–6 mmHg, the P_{CO_2} was 92–110 mmHg and the pH was 6.88–6.96.

Allen (1937, 1938a, b) has reported the survival of limbs of dogs, rats, rabbits and cats after arrest of the circulation for 15 h. He concluded:

> There is no theoretical reason why human limbs should be any less resistant. On the contrary, it is probable that the activity of tissue metabolism is one of the important factors determining the survival limit, and because of the rapidity of metabolism this limit is likely to be shorter in the mouse, rat, and chicken than in the dog, and perhaps shorter in the dog than in man. A few isolated examples indicate that human limbs are at least as resistant as those of animals (Allen, 1938b).

Paletta, Willman & Ship (1960) examined the effects of applying a tourniquet to the hindlimb of dogs for periods of up to 5 h. Oedema at the site of the tourniquet resolved in one week, but foot drop persisted for two weeks. Thereafter, the animals were apparently normal, but histological examination of limb tissue revealed neuromuscular damage (Walker, Paletta & Cooper, 1959). Rabbit skeletal muscle has been shown to contract on direct stimulation after 3 h of ischaemia, and nerve conduction was still present after 6 h of ischaemia (Holubár, Schück & Saravec, 1952).

Circulatory control of metabolism? Its possible role in asphyxial defence

The basic priority of the circulation is that blood flow should be supplied to tissues in accordance with their needs. This primary role leads one to think of the circulation as the servant of tissue metabolism. The circulation may also act as a controller of metabolic rate, as can be demonstrated in limbs.

The effect of decreased blood flow on tissue metabolism has been studied in skeletal muscle. Pappenheimer (1941) found that at flow rates greater than 40 ml/min in a 10 kg dog, the rate of oxygen consumption in the isolated perfused

hindlimb of the dog was independent of blood flow. However, below this value the oxygen consumption decreased,
reaching 50% of its normal value when the blood flow was
reduced to about 10 ml/min. Fales, Heisey & Zierler (1962)
studied the dog gastrocnemius-plantaris muscle *in situ* during
partial venous occlusion and compared this procedure with
the effects of arterial occlusion. They found that, within
limits, muscle oxygen consumption was dependent on or
limited by blood flow. However, Stainsby & Otis (1964) did
not find such a relation except at very low levels of flow.
Whalen, Buerk & Thuning (1973) suspected that the discord-
ant results might be attributed to differences between red and
white muscle. Caldwell & Wittenberg (1974) studied the
relation between oxygen consumption and oxygen pressure
in tissue slices from rat kidney, heart, liver, brain, diaphragm
and lung. The results demonstrated that the oxygen utilization decreased progressively below an oxygen tension of 100
mmHg for all of the tissues tested.

Nakamura (1921) stimulated the lumbar sympathetic
nerves in the cat and found a pronounced decrease in both the
rate of blood flow and the oxygen consumption in the lower
leg. Rein & Schneider (1937) reported that vasoconstriction
in the dog's hindlimb reduced the oxygen consumption of the
limb, and the arteriovenous temperature difference was also
diminished. The magnitude of the latter decrease was of the
order which might be expected from the calculated decrease
in oxygen consumption.

The priority of asphyxial defence

Where does the asphyxial defence response stand in relation
to other circulatory functions? This question of the functional priorities of the circulation may be answered, at least
partially, by examining situations in which an asphyxial
threat is combined with another challenge which by itself
would normally claim circulatory precedence. Human experiments using face immersion with breath-holding as the
asphyxial threat have been performed simultaneously with
one of three other stresses: local tissue ischaemia, exercise and
body heating. Face immersion with breath-holding evokes

49

bradycardia and decreased peripheral blood flow whereas the other three conditions produce hyperaemia.

In order to determine whether the vasoconstriction evoked by diving in man could depress reactive hyperaemia pro-, duced by the release of limb circulatory arrest, the following laboratory experiment was performed (Elsner & Gooden, 1970). Forearm blood flow was measured by plethysmography in eight subjects. The flow was arrested by a high pressure cuff on the upper arm for periods of 2, 5 and 10 min, and upon release of the cuff the resulting reactive hyperaemia was measured. The procedures were repeated with the subjects performing a 1-min face immersion in water, which began 30 s before the release of the arresting cuff and ended 30 s after release of the cuff. As a result of the superimposition of the face immersion, the volume of the reactive hyperaemia after all of the three arrest periods was significantly reduced (Fig. 3.7). After 2 and 5 min of arrest the volume was 46% and 45% respectively of the hyperaemia following arrest alone. The reduction in the hyperaemic flow was particularly striking in some subjects.

The persistence of diving bradycardia during active underwater swimming has been observed in man by several workers (Wyss, 1956b; Olsen, Fanestil & Scholander, 1962a; Scholander et al., 1962a; Irving, 1963; Hong et al., 1967; Strømme, Kerem & Elsner, 1970). Asmussen & Kristiansson (1968) found reduced heart rate in response to face immersion during exercise on a bicycle ergometer. In one subject the rate decreased to 10 beats/min during exercise at 800 kpm/min. Strømme et al. (1970) studied the heart rate responses of 40 subjects, males and females, during rest and swimming exercise combined with face immersion breath-holding. Despite an elevated heart rate during exercise experiments, face immersion reduced the rate to a level below that produced under resting conditions, in one subject to a pulse interval of 6 s. The results of this study suggest that the combination of exercise and face immersion breath-holding can result in an even stronger stimulus for bradycardia than face immersion breath-holding at rest. Strømme & Blix (1976) proposed that this intensification of diving bradycardia is due to chemoreceptor reflex facilitation.

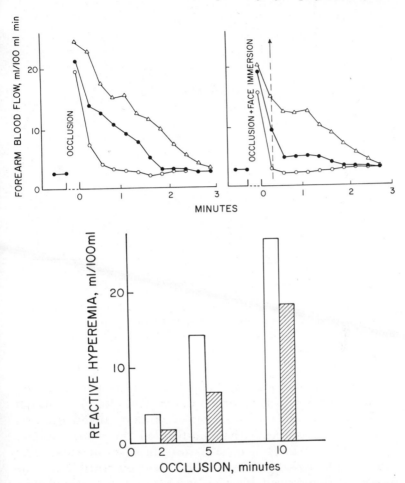

Fig. 3.7. Average blood flow of eight human subjects after 2 (open circles), 5 (filled circles) and 10 (triangles) min of occlusion with and without face immersion (above). Cumulative reactive hyperaemia volume (below). Open columns, without face immersion; cross-hatched columns, with face immersion. (Redrawn from Elsner & Gooden, 1970.)

To determine whether the sympathetic vasoconstriction induced by diving in man can suppress exercise hyperaemia, three subjects who showed a particularly pronounced reduction in calf blood flow during an earlier face immersion

51

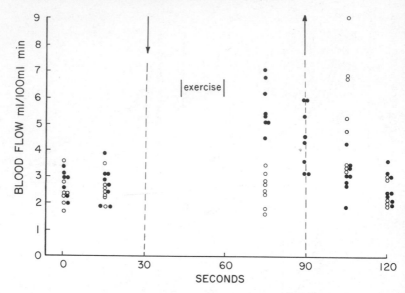

Fig. 3.8. Calf blood flow in three selected human subjects showing post-exercise hyperaemia with (open circles) and without (filled circles) face immersion. Beginning and end of face immersion indicated by arrows. (Redrawn from Elsner *et al.*, 1966*a*.)

experiment were tested in a combined study (Elsner *et al.*, 1966*a*). Calf flow was recorded before and after 15 s of foot pedal exercise, and the usual hyperaemia occurred after the exercise. The subjects then performed the same exercise during the first 30 s of a 1-min face immersion. The subsequent hyperaemia was then delayed until after the subject had removed his face from the water. Blood flow measurements taken immediately after exercise, but still during the face immersion, did not show the customary post-exercise hyperaemia. Thus the locally mediated exercise hyperaemia was suppressed, presumably by the sympathetic vasoconstrictor response to face immersion (Fig. 3.8).

Studies on rat and man have examined the influences of face immersion with breath-holding on limb blood flow that was already elevated in response to an external heat load. Lightly anaesthetized rats were subjected to warming that produced a persistent vasodilation in the tail (Gooden, 1971*a*). Blood flow through the tail was measured by venous occlusion plethysmography. The rat's head was immersed in water at 20°C for durations of up to 40 s. This procedure caused a rapid and pronounced reduction in tail flow, often to zero, within 15–30 s. The subcutaneous tail temperature fell throughout the immersion and recovered with subsequent recovery of the tail flow. It seemed unlikely that the tail flow response was simply the result of stimulation of temperature receptors on the face since thoroughly moistening the face with water in the absence of apnoea caused a transient increase in tail flow.

A similar experiment was performed on human subjects (B. Dale, S. Ireland, P. Kanas, M. Lindsell, P. Randall & V. Thaller, personal communication). The response of hand flow to face immersion was recorded in volunteers at normal and elevated body temperature (increased by 1°C). At normal temperature face immersion produced a small reduction in hand blood flow. Face immersion at elevated temperature produced a marked reduction in hand flow which fell initially to one-third of the heated control level. These experiments on rats and men suggest that the sympathetic vasoconstriction of the diving response suppressed the vasodilation of the thermoregulatory response. Both vascular responses are centrally mediated, and the asphyxial defence mechanism apparently took precedence over thermoregulation.

Experiments on harbour seals by Elsner *et al.* (1975) and Hammel *et al.* (1977) suggest that under some conditions the need to lose heat may be overridden by the circulatory responses of diving such as decreased peripheral blood flow. Thermodes, through which heating or cooling water could be circulated, were implanted into the preoptic hypothalamic region. In this manner the temperature of the central thermoregulator could be manipulated as a means of controlling thermoregulatory responses. In some experiments superficial vasodilation induced by hypothalamic heating was promptly inhibited at the onset of an experimental dive and was reactivated again after the dive.

Biochemical adaptations

Attempts to discover enzymatic and biochemical processes which might endow diving animals with special adaptations for long periods of asphyxia have been inconclusive and difficult to interpret. Possible mechanisms for the support of anaerobic glycolysis in hypoxic tissues and for the efficient use and conservation of oxygen as well as recovery adjustments have been examined. They have been the subject of a recent review (Kooyman, Castellini & Davis, 1981). The respiratory quotient of resting seals generally falls in the range which indicates that fat is the primary metabolic fuel, and carbohydrate usually provides 10% or less of aerobic energy metabolism.

The use of radioisotope turnover techniques has yielded new information concerning metabolism during diving. Anaerobic glycolysis in diving harbour seals accounted for about 15% of resting metabolism in 10-min dives (maximum diving time about 25 min). Both blood glucose and tissue glycogen were metabolized. Most of the lactate was recycled during post-dive recovery, and about 30% was oxidized (Kooyman *et al.*, 1981).

When the occasion arises for diving seals to resort to anaerobic metabolism, some species have abundant stores of glycogen available for fuelling that process. Glycogen storage in the cardiac muscle of Weddell seals (Kerem, Hammond & Elsner, 1973) and harbour seals (F. White & R. Elsner, unpublished observations) is two to three times as great as in terrestrial mammals. In vitro experiments indicate that the ischaemic harbour seal heart consumes its glycogen (F. White & R. Elsner, unpublished observations) and skeletal muscle utilizes glycogen during dives (Scholander *et al.*, 1942*a*). Brain tissue ordinarily contains little glycogen. However, Weddell seal brain has a higher content of that substance than dog, cat, rabbit or rat brains (Kerem *et al.*, 1973). Seal heart muscle is also unusually rich in myoglobin (Blessing & Hartschen-Niemeyer, 1969). Some information concerning cardiac metabolism during long seal dives has come from recent studies involving catheterizations which permit sampling of blood from the coronary sinus and the arterial circulation. When this was combined with myocardial blood flow data from the radioactive microsphere technique, indications of substrate consumption and metabolite production were obtained. Free fatty acids, the generally preferred oxidative fuel (Neely & Morgan, 1974), continued to be extracted during diving. Paradoxically, however, lactate began to be produced by the heart even when there still remained a substantial oxygen content in the arterial blood. As noted earlier, oxygen consumption steadily decreased. Myocardial arteriovenous glucose differences declined early in the dive and remained depressed until the post-dive recovery (Kjekshus *et al.*, 1982).

These puzzling results, which suggested simultaneous aerobic and anaerobic cardiac metabolism, seemed incompatible until re-examination of myocardial blood flow with chronically implanted coronary flowmeters revealed its intermittent nature (Fig. 3.4). An alternation of anaerobic glycolysis and oxidative metabolism associated with the oscillating perfusion suggests an effective means for extending the available oxygen reserves while at the same time sparing blood glucose for the brain and preventing accumulation of certain end-metabolites in myocardial tissue.

Lactate production, and by inference anaerobic glycolysis, has been demonstrated to develop progressively in harbour seal brain as oxygen consumption declines during lengthy laboratory dives (Kerem & Elsner, 1973*b*). Such a conversion to anaerobic processes would be unusual in brain tissue of terrestrial mammals, which usually depends upon oxidative metabolism. Diving ducks also depend exclusively upon oxidative brain metabolism and a continuous oxygen supply to the central nervous system maintained by appropriate circulatory distribution (Bryan & Jones, 1980*a*, *b*). Murphy, Zapol & Hochachka (1980) suggested that differing substrate preferences in Weddell seal tissues help to assure that limited energy sources will be divided among them. In that study the high blood concentration of lactate flushed from previously under-perfused tissues during post-dive recovery appeared to be readily metabolized by brain, heart and lung.

Reviews of efforts to discover specific adaptations in enzyme activities which might be related to enhanced glycolytic capability in diving animals have appeared (Hochachka & Storey, 1975; Blix, 1976; Hochachka & Murphy, 1979; Hochachka, 1980; Castellini, Somero & Kooyman, 1981; Kooyman *et al.*, 1981). The evidence is confusing and

55

sometimes contradictory. Three marine mammal species (sea lion, harbour seal and Weddell seal) were compared by Simon *et al.* (1974) for relative activities of pyruvate kinase, a glycolytic regulatory enzyme, from heart, brain, skeletal muscle, lung and liver. While activities correlated positively with maximum diving times (longest in Weddell seals, shortest in sea lions), the significance of that observation is in some doubt, since pyruvate kinase is similarly active in the tissues of several terrestrial species. Dog and harbour seal cerebral hexokinase and lactate dehydrogenase (LDH) are essentially alike (Jarrell, 1979), but the Weddell seal cardiac LDH level is reported to be the highest of any mammalian heart studied (Hochachka & Murphy, 1979). Castellini *et al.* (1981) and Kooyman *et al.* (1981) marshal arguments to defend the proposition that major adaptations of marine mammal enzyme activities in the service of breath-hold diving have not been demonstrated, since comparisons with terrestrial species have not revealed substantial and convincing differences.

A somewhat different approach has been taken in studies of enzyme properties which would be expected to favour the continued operation of metabolic activity during prolonged diving asphyxia. Fructose 1,6- bisphosphatase (FbPase) (which plays a role in the regulation of gluconeogenesis, expected to be active during recovery) from dog and seal liver and kidney were compared. Activities at pH 7.0 were not substantially different, but seal FbPase was markedly less sensitive to pH change than the dog enzyme over a range from pH 6.8 to pH 7.4, suggesting another potential adaptive mechanism (Behrisch & Elsner, 1980*a*). Enzymatic adaptation may in this instance take the form of lessened sensitivity to pH which would allow reaction catalysis to proceed in specific seal tissues while it would be inhibited by acidosis in similar tissues of terrestrial species. Rapid post-dive restoration of available glucose is suggested by elevated plasma glucagon in harbour seals (Robin *et al.*, 1981).

Another study suggests that glycerol, in addition to glucose, can participate as an anaerobic glycolytic fuel in ischaemic seal kidneys (Behrisch and Elsner, 1980*b*). When perfusion in isolated kidneys was arrested, glycogen within the cortex was converted stoichiometrically to lactate. After glycogen was finally depleted, lactate production continued and was accompanied by an increase in free fatty acids but without accumulation of glycerol. Appearance of this mechanism after long ischaemia suggests that it constitutes an emergency reserve. Despite the tissue's low energy charge by that time, indicating much-reduced ATP availability, restarting perfusion resulted in restoration of oxidative activity. Tissue studies which demonstrate the seal kidney's anaerobic capacity have been referred to earlier in this chapter.

Comparisons of enzyme activities from marine mammal tissues with those from terrestrial, presumably 'non-diving', species is a doubtful and possibly not very instructive procedure. For instance, glycolytic enzymes in the muscles of seals, dogs, rabbits and rats can be expected to serve biological functions having very considerable interspecies variability and, therefore, little relation to diving adaptation. An alternative description of enzyme characteristics as a ratio of activities in an organ which is ischaemic during dives, for example seal kidney, and one which is relatively well perfused, for example heart, may provide a more useful comparison of species. Thus the kidney : heart ratios for enzymes catalyzing the anaerobic metabolism of glycerol were much higher for seal than for dog tissues (Behrisch & Elsner, 1980*b*).

Ability to perform muscular exercise, as measured by maximum oxygen consumption, is roughly eight times greater in dogs (Langman, Bandinette & Taylor, 1981) than in harbour seals (Ashwell-Erickson & Elsner, 1981), suggesting that the seal is relatively sedentary while the dog is an elite athlete. The recent discovery that harbour seal heart has less mitochondrial cytochrome and respiratory enzyme activity per unit of tissue protein than does dog heart (Sordahl *et al.*, 1981; Sordahl, Mueller & Elsner, 1982) is likely to be related to the seal's unimpressive maximum oxygen consumption and may have less relevance to its diving abilities. This characteristic exercise performance also allows speculation concerning an evolutionary trade-off in seals which favours a large blood oxygen storage, and its related high haematocrit, at the expense of reduced circulatory gas transport function resulting from the associated high blood viscosity.

Despite the increased buffering capacity with which seal

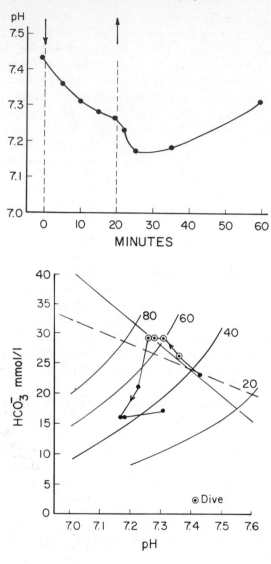

Fig. 3.9. Changes in arterial blood pH during and after an experimental dive in an elephant seal *Mirounga angustirostris* (above). Bicarbonate–pH diagram of acid-base status of arterial blood from the same samples (below). Data obtained during the dive indicated by filled circle within an open circle. Carbon dioxide isobars labelled 20, 40, 60, 80 (mmHg).

blood is endowed, mainly by haemoglobin, a relentless acidosis nevertheless occurs during long dives. There is often a further fall in pH after the dive when lactic acid is flushed from the tissues. Fig. 3.9 shows the changes in arterial blood pH during one such dive of an elephant seal. Also shown is a conventional acid–base diagram, the Davenport (1974) bicarbonate–pH plot, for the same dive. The data indicate an increasing respiratory acidosis that advances slightly above the normal buffer curve as the dive progresses and haemoglobin is reduced, then changes to metabolic acidosis and returns towards normal conditions during recovery. Similar data, but showing increased metabolic acidosis, were obtained in experiments on ducks (Andersen, Hustvedt & Lövö (1965). The contribution of lactic acid to the metabolic acidosis of the seal after the dive was determined by comparing its concentration with 'fixed acid' production (Davenport, 1974). The average fraction of lactic acid thus estimated was 73% of the total fixed acid during recovery from diving in the five seals studied (R. Elsner, P. F. Scholander & A. B. Graig, unpublished results). Comparable acid–base alterations would constitute a severe disruption of body chemistry for most terrestrial mammals. An interesting question can be posed concerning the possible interplay of increasing hydrogen ion concentration, which tends to inhibit metabolism (Harken, 1976), and the seal's relative tolerance of acid metabolic products which accumulate in its tissues during diving.

Caption for fig. 3.9 *(cont)*

Normal buffer curves of elephant seal blood, solid line (Lenfant, 1969) and human blood, dashed line (Davenport, 1974). R. Elsner, P. F. Scholander & A. B. Craig, unpublished results.)

4 HUMAN DIVERS

In the type of free dives employed by the Ama the engineering problems appear to be rather simple, and involve chiefly the method of getting the diver up and down quickly so that the time on the bottom can be as long as possible. Most of the problems, however, will always be physiological, and they deal with human responses at the very margin of tolerance limits. It is, therefore, a fascinating situation that merits thorough investigation (Wallace Fenn, 1965).

Historical background

Early man was a nomadic food gatherer. He took advantage of almost any available food source and this included sea food. At a recent excavation of a *Homo erectus* site at Terra Amata on an ancient Mediterranean seashore archaeologists uncovered the shells of oysters, mussels and limpets as well as fish-bones (de Lumley, 1969). These remains show that the ancestors of modern man were using the sea as a source of food more than 40 000 years ago. Some shells found in prehistoric middens are of varieties that grow in relatively deep water, and this suggests that they were collected during breath-hold diving (Nukada, 1965).

Initially, diving was probably performed to supplement other food supplies. As time passed sea food became increasingly important as an item of trade with other groups. Products such as pearls and sponges were sought for reasons other than nutrition. The abalone were used in Japan for aesthetic as well as religious purposes. Breath-hold diving for mother-of-pearl in the Mediterranean is reported to have occurred in about 3000 BC with dives to depths of 50 feet (15.2 m).

An early record of breath-hold diving occurs in the Iliad where the fall of Hector's charioteer is compared with the

60

action of a diver in search of oysters (Cross, 1965). Divers were not only interested in bringing the natural flora and fauna to the surface. As the civilized world became more technically advanced they also sought man-made objects on the ocean floor. In 460 BC Herodotus wrote of a Greek diver employed to recover treasure from sunken ships. Naval divers were trained and employed as part of military forces as early as 333 BC.

Some of the better-known fisheries that employed breath-hold diving may have had their origins in prehistory. These include the Mediterranean sponge and Ceylonese pearl fisheries. Spanish writers of the sixteenth century indicate that there was extensive employment of divers in the Gulf of Mexico at that time. Pearl diving also still occurs in the Tuamotu Archipelago, and trochus-shell diving is pursued in the region of Thursday Island off the coast of north-eastern Australia.

Undoubtedly the most colourful and well-documented groups of human breath-hold divers are the Ama of Japan and Korea. Detailed reviews of their activities are found in *Physiology of Breath-hold Diving and The Ama of Japan* edited by Rahn and Yokoyama (1965). In the old Japanese language the word Ama originally meant the ocean or sky. Many ancient references indicate that this diving activity has existed for at least 2000 years. One of the older references is the Gishi-Wajin-Den which is believed to have been published in 268 BC (Nukada, 1965). The Ama may be either male or female but ordinarily the word is applied to female divers. These women breath-hold dive for the pearl culture industry or to collect marine plants and animals fixed to reefs, and they live partially or completely on the income from these harvests (Kita, 1965). Along the shores of the Korean peninsula are found the Korean Ama or Hae-Nyo. Hong and Rahn (1967) stated that some 30 000 Ama were working along the sea coasts of Japan and Korea. Both the Japanese and Korean groups are decreasing in numbers, and Hong (1965) estimates that the Korean Ama will have ceased their work by the year 2000.

The Japanese Ama may be divided into three grades (Nukada, 1965). The Koisodo are usually young girls in training. They dive to only 2–4 m and stay close to shore. The Nakaisodo are girls who have already been trained as Koisodo for a few years. They are 15–20 years old and dive to 4–7 m. The most capable Ama are called Ooisodo. They are usually more than 20 years old and dive to 10–25 m. They descend rapidly with a counterweight and are subsequently pulled to the surface by an assistant, usually a man, in a boat. Each dive lasts approximately 40–50 s with a 50- to 60-s rest period between dives. The Ama perform about 50 dives in the morning and a similar number in the afternoon.

Many suggestions have been given to explain why men do not engage in this type of diving. Hill (1912) recounted the following anecdote.

> A distinguished Japanese artist tells me that as a child he heard the origin of this custom lay in the fact that the very cold water in the depths had a perishing effect on the testicles of the men. It was reputed that the sea-bream attacked the men and bit off their testicles, but it was really the cold. The women having no vital parts so exposed to cold, took over the work, both for the men's sake and their own.

Nukada (1965) suggests that the men were engaged in off-shore fishing or as sailors on ships. The Ama stayed home and took up diving to supplement the farming harvest from the land. The deep-diving Ooisodo require a boatman and this assistant is usually the father, husband or brother of the diver. Korean women tested by Rennie *et al.* (1962) had significantly higher basal metabolic rates and greater body insulation from subcutaneous fat than did Korean men.

Maximum diving ability

This field is notorious for fraudulent claims, and the length of time that a man can remain under water without breathing has often been greatly exaggerated. Robert Boyle (quoted by Hill, 1912) was aware of this in the seventeenth century when he wrote:

> Those that dive for pearls in the West Indies are said to be able to stay a whole hour under water; and Cardan tells us of one Calamus, a diver in Sicily, who was able to continue

(if Cardan neither mistake nor impose upon us) three or four times as long.

However, Boyle questioned the veracity of these stories and continues:

> An ingenious man of my acquaintance, who is very famous for the useful skill of drawing goods, and even ordnance, out of sunk ships, being asked by me how long he was able to continue at a depth of 50 to 60 feet under water without the use of respiration, confessed to me that he cannot continue above two minutes of an hour without resorting to air, which he carries down with him in a certain engine . . .

Rahn (1964) calculated that the total oxygen store in a 70-kg man at resting lung volume was approximately 1.5 l. Breath-holding, asphyxia or submersion prevents replenishment of the oxygen stores, but experiments on dogs suggest that nearly all of the oxygen can be extracted from the stores before death (Herber, 1948). One might therefore predict that a resting man who has an oxygen consumption of 0.3 l/min would completely deplete his oxygen stores in 5 min. However, a man at rest without prior hyperventilation and starting with a resting lung volume can hold his breath for approximately 1 min.

The world record for breath-holding underwater is 13 min 42.5 s and was achieved by R. Foster at the age of 32. He remained submerged under 3 m of water in a swimming pool of the Bermuda Palms Hotel at San Rafael, California, on 15 March 1959. Before his record dive he hyperventilated on oxygen for 30 min and then lay relaxed on the pool bottom with weights to prevent him from floating to the surface. He wore a complete wet suit including hood and face mask. His longest breath-hold without oxygen was 5 min 40 s (McWhirter & McWhirter, 1973; McWhirter, 1980). J. Mayol holds the record for the deepest breath-hold dive, 100 m (cited by Hempleman & Lockwood, 1978).

Morphological and physiological adaptations

Adaptations that occur in human breath-hold divers encompass both morphological and physiological changes. Several anthropometric studies of the Ama have been undertaken (see review, Tatai & Tatai, 1965). Measurements such as weight, upper arm circumference, upper arm and thigh skin-fold thickness, grip strength and back muscle strength are all significantly higher in Ama than in farm women living in the same village. However, the most conspicuous feature of the physical characteristics of the Ama is their superior ventilatory function. Their vital capacity and maximum breathing capacity are much larger than those of non-divers. The increase in vital capacity is due to an increase in the inspiratory reserve volume (Song *et al.*, 1963).

Teruoka's paper entitled 'Die Ama und ihre Arbeit', published in 1932, is accorded the distinction of being the first scientific treatise dealing with the physiology of breath-hold diving (Rahn, 1965). He described the patterns, frequency, depth and duration of Ama dives and how they vary with water temperature and the seasons. He analyzed alveolar gas exchange by collecting the first exhalation after dives as deep as 25 m and lasting as long as 118 s. It was not until 1959, 27 years after the publication of Teruoka's paper, that S. Hong at Yonsei University College of Medicine, Seoul, Korea, initiated a programme of further research into the Korean Ama.

The limit, or breaking point, of breath-holding is determined by the interactions of lung volume, carbon dioxide and oxygen blood gas tensions as well as psychological drive (Mithoefer, 1965; Godfrey & Campbell, 1968). The greater than normal vital capacity of the Ama would be expected to permit a longer duration of breath-holding. Diving women develop a smaller increase in their ventilatory response when exposed to hypercapnic gas mixtures than do non-diving laboratory personnel (Song *et al.*, 1963). This difference in the response of trained divers may result from a decreased sensitivity of the medullary centres to carbon dioxide or a conscious disregard of the carbon dioxide stimulus, or both (Goff & Bartlett, 1957). The Ama and the non-divers showed no significant difference in their ventilatory responses to a hypoxic gas mixture. Similarly, Schaefer (1955) found a depressed ventilatory response to carbon dioxide in instructors who worked regularly in a submarine escape training tank compared with that of laboratory personnel.

An elevation of circulating catecholamines has been shown to increase the respiratory sensitivity of human subjects to hypoxia and hypercapnia (Cunningham *et al.*, 1963). People who dive regularly appear to develop less stress reaction before and during a dive than inexperienced divers (Davis *et al.*, 1972; Gooden, Feinstein & Skutt, 1975). A smaller increase in circulating catecholamines in regular divers might therefore be expected to result in a relative reduction in sensitivity to both hypercapnia and hypoxia.

Cardiovascular responses to diving

Perhaps one of the earliest allusions to a cardiovascular response to breath-hold diving in man is found in a book of travels in the Northern Sahara by Colemieu cited by Heller, Mager & von Schrötter (1900). The author describes a group of desert dwellers who dive for the purpose of maintaining their supply of fresh water. In this account the author mentions a diver's pulse rate before and immediately after diving.

The water-supply is obtained from artesian wells which are up to 130 feet in depth. The 'eye' of these wells often gets stopped up with sand, brought up from below by the spring or blown in from above. The difficult and dangerous work of clearing the 'eye' is carried out by a guild of divers. A single diver brings up as much sand as he can in three to four minutes. Not infrequently a diver loses his life over his work. The descent and ascent are made by means of ropes which are let down the well. The diver stops up his ears with wax, and, getting into the water, waits a while to get used to the temperature. He then gives the signal, fills his lungs as much as he can with a couple of breaths, and sinks under. His comrades can follow his movements by the two ropes which reach to the bottom. Three minutes and two seconds have already gone by, and one is beginning to get anxious, when he rises to the surface half asphyxiated and almost unconscious. His comrades grip him and hold him while he gets his breath, then he climbs out and goes and warms himself by the fire, while the next takes his turn.

65

The young men of the guild seem strong and healthy, the old lean and straight-chested, but they stay under longer and suffer less. The young are too hasty, and that tells against them. The pulse is noted as being diminished in frequency – e.g., 86 before the descent, and 55 immediately after the ascent.

Human experiments in which diving bradycardia was recorded, even during underwater swimming, first appeared in 1940. These were mentioned briefly in a paper describing the respiratory metabolism of the porpoise (Irving, Scholander & Grinnell, 1941*a*), but were not fully reported until 1963 when they became of historical importance (Irving, 1963). For the 20 years following this brief report in 1940, the field lay virtually dormant apart from two papers by Wyss (1956*a*, *b*) in which diving bradycardia and cardiac arrhythmias were reported.

Bradycardia has been observed in human subjects during actual and simulated diving. In one particularly striking case Elsner *et al.* (1966*a*) observed a heart rate of 13 beats/min in a subject performing a 30-s face immersion. This subject began the procedure with a heart rate of 90 beats/min. The record appears to be a pulse interval equivalent to 7.3 beats/min during face immersion in cold water (R. Arnold, S. Krejci & R. Elsner, unpublished results).

In 1961 Scholander led an expedition to the Torres Strait off north-eastern Australia for the purpose of studying at first hand the physiological responses that occur in native skin-divers during breath-hold dives (Scholander *et al.*, 1962*a*). All of the 31 divers who took part in these studies were natives from the Thursday Island area and the adjacent coast of Cape York Peninsula. These men had been skin-diving for trochus shells among the islands and along the Great Barrier Reef since boyhood and earned their living by this means. The men used only a conventional face mask and dived for these shells in the shallow waters of the coral reefs.

The investigations concentrated on three topics: heart rate, arterial blood pressure and blood lactic acid level. Electrocardiograms were obtained during dives to depths of 6–8 m and lasting from 0.75 to 1 min. In every case the heart rate increased slightly at the start of the dive but then slowed to about 60% of the pre-dive rate. The reduction was commonly to 50% of the pre-dive value and sometimes even lower in spite of the underwater exercise. In the recovery phase the normal rate was restored rapidly. Arrhythmias developed in some divers during the dive and sometimes persisted into the beginning of the recovery period. In five divers who were subjects for this experiment, there was a transitory rise in venous blood lactate during recovery.

Sasamoto (1965) monitored the electrocardiograms of Japanese Ama during dives to 10 m. Heart rate slowed below the resting level within 20 s of diving. Ectopic beats were noted late in the dives. At about the same time Hong *et al.* (1967) were investigating the cardiac responses of the Korean Ama. They observed a greater diving bradycardia in these women during winter dives than during summer dives. Cardiac arrhythmias were common during breath-holding in air and upon submersion. The incidence was approximately 43% in summer and 72% in winter. Abnormal P wave and atrioventricular (A-V) nodal rhythms occurred most frequently.

The cardiac rhythm was studied during experimental dives by experienced divers (Olsen *et al.*, 1962*a*). Cardiac arrhythmias occurred in association with 45 of 64 periods of apnoea and were more frequent during total immersion than during simple breath-holding in air. The more severe arrhythmias generally occurred towards the end of apnoea. Atrial, nodal and ventricular premature contractions were observed during apnoea and recovery. The other arrhythmias were all of the inhibitory type, and included sinus bradycardia, sinus arrest followed by either nodal or ventricular escape, A-V block, A-V nodal rhythm and idioventricular rhythm. The T waves of the electrocardiogram often become tall and peaked. Wyss (1956*a*, *b*) was the first person to observe rhythm changes in man during total immersion and he also noted high, pointed T waves.

Olsen, Fanestil & Scholander (1962*b*) recorded blood pressure via a Cournand needle in the brachial arteries. Their subjects hyperventilated before submersion and submerged after maximal inspiration. During some dives the subject

67

performed exercise by pushing against pedals attached by a fulcrum and pulleys to weights. The arterial blood samples were analyzed for pH, carbon dioxide and oxygen content and lactate. Blood lactate increased by a small amount during the latter half of exercise dives, but the peak level was reached during the first minute of the recovery period. With the progressive decrease in heart rate during the dive there was a coincident increase in the arterial blood pressure which suggested that a marked degree of peripheral vasoconstriction also occurred during diving. It was concluded that the responses were similar to those reported in other diving mammals but were less pronounced. In retrospect, the use of a face mask and water at 37°C probably all militated against the development of a pronounced diving response in these experiments.

Elsner, Garey & Scholander (1963) evoked the diving response by simple face immersion with simultaneous breath-holding and found that this procedure was as effective in producing diving bradycardia as total body immersion. The subjects for this study were nine young men of varying diving experience. They lay prone and performed face immersion with breath-holding in a bowl of water for 1 min. Calf blood flow was measured by venous occlusion plethysmography using a mercury-in-rubber strain gauge (Barcroft & Swan, 1953; Whitney, 1953). The simulated dives were performed after moderate inspiration or expiration without hyperventilation. Blood flow measurements were obtained before and after face immersion as well as at the 15-, 30- and 45-s points during the procedure. The limb blood flow consistently decreased more during face immersion than during breath-holding alone (Elsner *et al.*, 1966a). However, Brick (1966) reported that the reduction in forearm blood flow during breath-holding in air was similar to that produced by face immersion with breath-holding.

The reduction in calf blood flow observed by Elsner *et al.* (1963) was assumed to be the result of vasoconstriction, but proof of this assumption depended upon the simultaneous measurement of the systemic arterial blood pressure and limb blood flow. Wolf, Schneider & Groover (1965) demonstrated a reduction in the skin temperature of the fingers with a concomitant increase in arterial blood pressure during simulated diving. This also suggested an increase in peripheral resistance.

Heistad, Abboud & Eckstein (1968) measured brachial arterial blood pressure directly by cannulation as well as finger and forearm blood flow by venous occlusion plethysmography during breath-holding with and without face immersion. During 30 s of face immersion breath-holding there was a progressive rise in mean arterial blood pressure to an average of 20% above the control level. Both systolic and diastolic pressures were elevated. At the same time the finger and forearm blood flow decreased by an average of 72 and 50% respectively below the control value. Heistad *et al.* concluded that vasoconstriction must occur in the forearm vasculature during this form of simulated diving. Breath-holding alone produced similar but consistently smaller changes in pressure and flow.

Harding, Roman & Whelan (1965) investigated the cardiovascular response to breath-holding with the subjects standing in water at different depths including total immersion. This initial study led to an investigation of the cardiovascular responses to total immersion and the etiology of these responses (Campbell, Gooden & Horowitz, 1969*a*; Campbell *et al.*, 1969*b*). Breath-holding with the body supine and immersed except for the face produced a small increase in forearm vascular resistance, but during total immersion forearm resistance increased by 70%. The calf vascular resistance increased by 147%. During a profound individual response to total immersion, the forearm vascular resistance increased approximately 16 times (Fig. 4.1).

As in seals and ducks, vascular constriction during diving in man may not be limited to the arterioles. Olsen *et al.* (1962*b*) reported frequent damping of the brachial arterial pressure recording during total immersion of their diver subjects. They suggested that this effect resulted from the constricted arterial wall occluding the tip of the Cournand needle. Elsner, Gooden & Robinson (1971*a*) found considerable difficulty in withdrawing brachial arterial blood samples during face immersion breath-holding in a subject who had developed a profound reduction in forearm blood

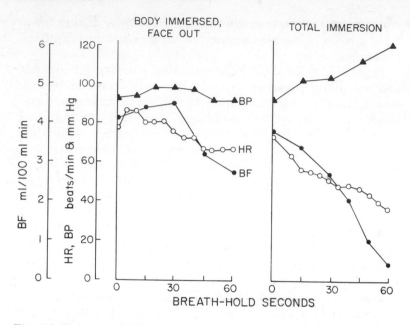

Fig. 4.1 Mean arterial blood pressure (BP), heart rate (HR) and forearm blood flow (BF) in a human subject during 1-min breath-holds. (Redrawn from Campbell *et al.*, 1969*b*.)

flow. Additional evidence that vasoconstriction occurs in human arteries of greater than arteriolar size during diving was obtained by Heistad *et al.* (1968). During simulated dives, a substantial systolic pressure gradient developed between the brachial and digital arteries in three out of five subjects, suggesting that vasoconstriction must have occurred in larger arteries upstream from the digital artery.

Heistad *et al.* (1968) also studied the reactions of capacity vessels during face immersion. By inflating a pressure cuff around the proximal part of a limb to 40 mmHg, venous return from the limb was prevented and the limb volume increased until it reached a relatively constant value. Any change in this volume is a qualitative estimate of a change in the volume of the capacity vessels, which are largely the veins (Wood, 1965). An abrupt and pronounced reduction in finger volume was noted during face immersion with breath-holding in seven subjects tested.

Kawakami, Natelson & DuBois (1967) calculated the cardiac index (cardiac output/m² of body surface area) using a nitrous oxide – plethysmographic method (DuBois & Marshall, 1957) during face immersion with breath-holding. They found an average of 2.5 l/min m², which was significantly less than the 3.2 l/min m² measured during breath-holding alone. This decrease in the cardiac index was due mainly to the bradycardia induced by face immersion rather than a decrease in stroke volume.

Functional significance of the diving response

The question arises whether the cardiovascular response to diving in man has any functional significance and in particular whether the human response has an oxygen-conserving role. Oxygen conservation could result in two ways. First, by limiting the blood flow to peripheral tissues such as the limbs and gut, the uptake of oxygen by these tissues would be reduced, and second, the reduction in the heart rate might decrease myocardial oxygen consumption.

When Olsen *et al.* (1962*b*) were studying blood gas tensions in their divers during total immersion, they noted a surprisingly high arterial blood oxygen content at the end of long exercise immersions. This observation suggested to them that there was a reduced extraction of oxygen in the tissues during diving in man. Wolf *et al.* (1965) monitored the change in arterial oxygen saturation by an ear oximeter during face immersion. In a harassed subject, who developed little or no diving bradycardia during face immersion, arterial oxygen saturation fell more rapidly than in a subject who developed marked bradycardia. Schaefer (1965) reported that professional divers breathing 10.5% oxygen in nitrogen developed a pronounced oxygen debt, several times greater than that of non-diving laboratory personnel. Schaefer ascribed this finding to a reduction in tissue oxidation in response to the hypoxia. Tibes & Stegemann (1969) reported a reduction in oxygen consumption in six male subjects during breath-holding in air and during diving.

Elsner *et al.* (1971*a*, *b*) examined the influence on the diving response of varying the arterial gas tensions of oxygen and

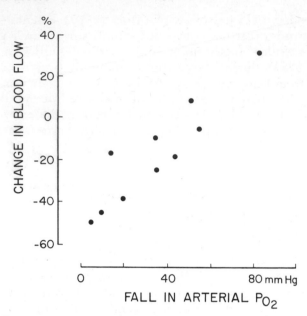

Fig. 4.2. Percentage change in forearm blood flow and decline in arterial oxygen tension during 5-min face immersions after oxygen hyperventilation in five human subjects. (Redrawn from Elsner *et al.*, 1971*a*.)

carbon dioxide before face immersion breath-holding. Brachial arterial blood was sampled from one arm, and forearm blood flow was measured by venous occlusion plethysmography simultaneously in the other arm. Five male subjects hyperventilated on oxygen for 15 min and then performed face immersions which lasted an average of 5 min. The mean arterial oxygen tension before face immersion was 548 mmHg, and after 4 min the tension was still considerably elevated in four of the five subjects (mean value, 326 mmHg). The remaining subject became anxious after 2 min of the procedure but voluntarily continued. His forearm blood flow increased by more than 50% above his resting control level instead of decreasing as in the other subjects. His arterial oxygen tension fell from 560 to 64 mmHg in 4 min.

A highly significant correlation existed between the change in forearm blood flow and the decrease in arterial oxygen tension (Fig. 4.2). Thus the greater the reduction in flow, the smaller was the decrease in oxygen tension. If this relation in the upper limb is representative of a more widespread phenomenon in the lower limbs, gut and elsewhere, then it seems likely that the diving response in man operates in some circumstances to conserve oxygen. Whatever practical importance may be attached to that is unknown. In contrast to these results, no reduction in oxygen consumption could be detected during human breath-holding face immersion in the following studies: Raper *et al.* (1967), Heistad & Wheeler (1970) and Hong *et al.* (1971).

5   CONTROL MECHANISMS

A living system is no more adequately characterized by an inventory of its material constituents . . . than the life of a city is described by the list of names and numbers in a telephone book. Only by virtue of their ordered interactions do molecules become partners in the living process . . . (Paul A. Weiss, 1969).

The rapid onset of respiratory and cardiovascular responses to diving and face immersion argues for their initiation by neural control mechanisms. Questions concerning these regulatory pathways have been subjected to study beginning with observations on diving ducks a century ago and continuing to the present day. It is possible to trace during this period four main lines of research into suspected reflexes which have led to the current state of understanding, often by routes in which the study of diving was not the primary motivation. These lines of research have involved reflexes originating from (1) receptors situated in the face, nasal mucosa and pharynx, (2) thoracic and pulmonary receptors, (3) peripheral arterial chemoreceptors and (4) arterial baroreceptors. More recently, the important role of the interactions of various respiratory and circulatory reflexes has come to be recognized.

Historically, much research into the nature of cardiovascular reflexes has depended upon experiments in which a discrete and isolated subsystem is stimulated and the unmodulated or passive output is recorded. The input of such an 'open-loop' reflex is independent of, and unaffected by, its output; it is deprived of feedback. Important contributions to our understanding of physiological controls have come from such experiments. However, the intact control mechanisms within the organism operate with information from a complex variety of output feedback or 'closed-loop' systems, the resultant outputs of which are powerfully modulated by interactions with related reflexes. Thus, while the reductionist approach of isolated and controlled experimentation has made possible the systematic description of individual reflexes, their study in re-assembled condition and their interactions with other reflexes provide insights, which are sometimes quite different, into the dynamic reactivity of these controls in their natural settings.

DIVING DUCKS AND SEALS

Recurring interest during the last 100 years of study has centred largely upon the neural controls governing respiratory and circulatory events in diving ducks, and systematic investigations of controls in other diving species were not performed until more recently. Some of the details of the responses in ducks which were matters of mild controversy have now been clarified. Differences in detail between the reactions of diving ducks and mammals, for example seals, have been the subject of recent studies. In experimentally dived ducks bradycardia develops gradually, requiring 30 s to 1 min for the full response. Comparable reactions are often initiated in 1 or 2 s in restrained and dived seals, more gradually in freely diving seals (Murdaugh, Seabury & Mitchell, 1961*b*; Elsner, 1965; Kooyman & Campbell, 1972; Jones *et al.*, 1973; Kooyman *et al.*, 1980).

The development of diving bradycardia and selective peripheral vasoconstriction is more complex than might be inferred from consideration of the few reflexes which collectively produce the diving response. The differences seen in the reactions to free diving when these are contrasted with observations of restrained animals in forced dives clearly indicate that the basic reflexes are subject to modification in both time and intensity by higher structures of the central nervous system, probably by conditioning and cortical influences. Forced diving appears to have little effect in manatees (Scholander & Irving, 1941), but profoundly intensifies the responses in seals (Scholander, 1940; Irving, Scholander & Grinnell, 1941*b*; Elsner *et al.*, 1966*a*). As noted earlier, attempts to force-dive dolphins *Tursiops truncatus* resulted in vigorous struggling, little heart rate slowing and sometimes death of the animal (Irving *et al.*, 1941*a*), while an animal of the same species trained to dive upon presentation of a target displayed a prompt and profound bradycardia (Elsner *et al.*, 1966*b*). Anaesthesia generally reduces or abolishes the apnoea, bradycardia and vasoconstriction, but techniques for maintaining some or most of the response even in anaesthetized animals have recently been developed (Daly, Elsner & Angell-James, 1977; Drummond & Jones, 1979).

Studies of influences, both external and internal, which combine to produce the diving response started with the duck experiments of Richet (1894). It was he who first suggested that the necessary condition for initiation of the diving response was that the duck be immersed in water and that the response could not be produced by asphyxia alone. Richet performed the experiment by closing the airways in two groups of ducks one of which was immersed in water and the other suffocated in air. The ducks immersed in water lived about three times as long as the ducks which remained in air. The results may be explained in part by the contrast in the ducks's reaction of relaxation to water immersion and its vigorous struggle following tracheal occlusion. Such relaxation is typical of natural divers and is undoubtedly a major reason for their economical consumption of metabolic reserves during submersion.

Richet was the first to show that diving bradycardia was mediated via the vagus nerve. It could be eliminated by division of the vagus or by administering atropine. The suggestion that the sensory source of information to the neural circuits involved might be triggered by impulses from the beak and head of the duck was first made by Richet (1894). The first experimental indication that this idea might be correct came from studies of the reflex nature of submersion apnoea and the discovery that the decerebrate duck was still capable of demonstrating the response of cardiac slowing upon immersion (Huxley, 1912, 1913*a*). The evidence from those experiments, therefore, indicated that the reflex pathway was independent of higher nervous centres. Lombroso (1913) was the first to attempt manipulation of the effects of water immersion and asphyxia. He showed that the submerged duck did not experience the usual diving bradycardia when the lungs were artificially ventilated.

The question of whether or not apnoea alone is a sufficient stimulus to produce bradycardia and other components of the diving response, or if other afferent inputs are required as well, has been a recurring theme in diving studies through the years. Huxley (1913*a*, *b*, *c*) reported that apnoea in ducks, whether produced by certain postures or by immersion of the head, resulted in heart slowing and that apnoea was the fundamental requirement for initiation of bradycardia.

The experiments of Andersen (1963*a*, *b*, *c*) and of Feigl & Folkow (1963) were similar in approach, both exploring the differences between submersion and asphyxia. When asphyxia was produced by plugging a tracheal cannula, the bradycardia which ensued was not as profound as that observed during head immersion. The heart rate returned to the pre-dive level immediately after breathing had restarted. Andersen showed that the application of water to the region of the nostrils was an adequate stimulus for the onset of apnoea and bradycardia and that the input was probably transmitted from sensory endings innervated by the trigeminal nerve. Furthermore, decerebrated ducks still showed apnoea and bradycardia when the head was immersed to the level of the nostrils, and these responses were reduced or eliminated when the trigeminal nerves were cut.

Natural mammalian divers also respond to trigeminal receptor stimulation. Application of water to facial surfaces innervated by the trigeminal nerve results in reflex apnoea in the expiratory position, bradycardia and peripheral vaso-constriction in seals (Scholander, 1940; Elsner *et al.*, 1966*a*; Dykes, 1974; Tanji, Weste & Dykes, 1975; Daly *et al.*, 1977; Elsner, Angell-James & Daly, 1977) and muskrats (Drummond & Jones, 1979). Similar responses result from laryngeal nerve stimulation in seals (Elsner *et al.*, 1977) and muskrats (Drummond & Jones, 1979).

Upper respiratory tract reflexes

It will be useful to consider a separate line of physiological investigation before returning to the topic of diving physi-ology. Experiments with various kinds of nasal and pharyngeal stimuli elicit cardiovascular responses relevant to this discussion. Kratschmer (1870) discovered that the unanaesthetized rabbit is highly sensitive to the inhalation of noxious gases: breathing smoke, for example, produced apnoea and bradycardia. These nasal reflexes have been the subject of continuing investigations, and their respiratory and circulatory effects have been described (Brodie & Russell, 1900; Dixon & Brodie, 1903; Allen, 1928; Ebbecke & Knüchel, 1943; Forster & Nyboer, 1955; Bynum, Ruoff & Rickert 1970; Angell-James & Daly, 1969b, 1972a; and White, McRitchie & Franklin, 1974).

The similarity of the responses to nasal stimulation and diving was recognized by Angell-James & Daly (1969a, 1972b), and they showed that the two phenomena are governed by the same underlying mechanisms. Vasomotor responses, which were elicited in lightly anaesthetized dogs by stimulation of the nasal mucosa, were found to include vasoconstriction in skin, muscle and splanchnic vascular beds which was the result of excitation of the sympathetic innervation. Blood flow was unchanged, or altered very little, in the carotid circulation. The responses were evoked by inducing a flow of water or saline through the nasopharynx and were abolished by cutting the trigeminal nerves or by local anaesthesia of the nasal mucosa. The cardiac response (bradycardia) was mediated by the vagus nerves.

Adrenal secretion of adrenaline and noradrenaline was increased during nasal stimulation by water or ether in rabbits. The increased output was abolished by anaesthesia of nasal passages. This and other evidence showed that the effect is reflex in nature (Allison & Powis, 1971). Similar reflex stimulation of catecholamine secretion has yet to be demonstrated in natural divers, but adrenal participation in the diving response is suggested by maintained blood flow in that gland during experimental dives in seals (Elsner *et al.*, 1978; Zapol *et al.*, 1979) and elevated blood corticosteroid levels in Weddell seals (Liggins *et al.*, 1979). Further evidence for a possible role of circulating catecholamines in sustaining vasoconstriction during long dives comes from studies of responses in isolated duck mesenteric arteries (Gooden, 1980b). Hypoxia was less effective in blocking the vaso-

78

constrictor response to noradrenaline than to nervous stimulation.

White (1975) and White, McRitchie & Korner (1975) showed the involvement of the trigeminal afferent input from the nasopharynx in detecting the presence of noxious gases in the inspired air and the resulting widespread cardiovascular events: bradycardia and reduced peripheral circulation, except for continued perfusion of the brain. They proposed that the circuitry governing these events was integrated via trigeminal nerves, a respiratory centre,* and vagal and vasomotor centres in the brain stem, and suggested that the cumulative result of the cardiovascular responses was a reduction in the rate of decline of the circulating blood oxygen content. The reflexes were, therefore, seen as protective by their resulting in lowered metabolism when the animal is threatened by potentially damaging environmental agents from which it might not immediately escape.

Sinus arrhythmia

Mechanisms controlling the variation of heart rate with respiration (sinus arrhythmia) are potentially related to the subject of controls operating during diving. The occurrence of sinus arrhythmia is well known; the increase in heart rate accompanies inspiration, and the rate slows again during expiration. The larger the inspiratory excursion, the greater will be the increase in the heart rate (reviewed by Daly, 1972). The explanation of the phenomenon lies in two likely mechanisms. One is a pulmonary stretch reflex, arising during inflation, which inhibits the cardio-inhibitory centre and thereby produces a decrease in normal vagal tone (Hering, 1871; Anrep, Pascual & Rössler, 1936a). Activation of this reflex also results in a slowing of respiration that is due to an increase in expiratory time, the classical Hering–Breuer inflation reflex. 'Irradiation' of impulses from the respiratory centres during inspiration to the cardio-inhibitory centre also causes its inhibition and results in tachycardia (Anrep,

---

* Use of the term 'centre' is not meant to imply exclusively a discrete nucleus of activity; it refers to a population of neurons serving a similar function.

Pascual & Rössler, 1936*b*). Cordier & Heymans (1935) further postulated that, during expiration, activity of the expiratory centre stimulates the cardio-inhibitory centre and thereby contributes to the heart slowing.

The relevance of this consideration of sinus arrhythmia to the present discussion rests in the observation of the enhanced influence of carotid chemoreceptor input upon vagally induced bradycardia when that input occurs during the expiratory phase of respiration (Elsner *et al.*, 1977*a*). This finding in seals is in agreement with the similar observations made on terrestrial mammals that chemoreceptor impulses impinging upon the respiratory centres during expiration produce a prolongation of expiration (Black & Torrance, 1971) and bradycardia (Haymet & McCloskey, 1975). Furthermore, when the carotid chemoreceptor stimulation was superimposed upon the trigeminal input, the resulting bradycardia was considerably enhanced to values exceeding the sum of carotid and trigeminal effects induced separately (Elsner *et al.*, 1977*a*).

Chemoreceptor reflexes

Yet another avenue of investigation has yielded important information for the understanding of neural controls and their interactions. Hollenberg & Uvnäs (1963) denervated the carotid chemoreceptors and baroreceptors of ducks and found that birds so treated and subsequently dived developed less than the usual bradycardia. Thus, progressive asphyxia appeared to act through the peripheral chemoreceptors in sustaining the diving responses. They argued that the effects seen were likely to be of carotid body chemoreceptor rather than carotid sinus origin because of the insensitivity of the baroreceptors to changes in blood gases and pH, but their results did not permit exclusion of a carotid sinus baroreceptor reflex as the explanation of the observed responses.

That the carotid bodies play a role in the diving responses of the duck was also demonstrated by Jones & Purves (1970) who denervated the carotid bodies of ducks and thereby drastically reduced their diving responses (Fig. 5.1). Holm & Sørensen (1972) obtained similar evidence. Butler & Jones

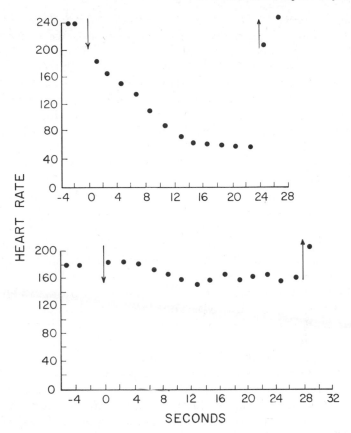

Fig. 5.1. Heart rates (beats/min) of diving ducks without (above) and with (below) carotid body denervation. Arrows indicated beginning and end of dive. (Redrawn from Jones & Purves, 1970.)

(1968) also showed that stimulation of the duck's glosso-pharyngeal nerve, which includes afferent input from the glottis, produced apnoea and bradycardia. This latter observation was confirmed by Jones & Purves (1970) and Blix (1975).

Daly and his co-workers showed that carotid body chemo-receptor stimulation in dogs produced a set of responses, including bradycardia and peripheral vasoconstriction, which closely resembled those observed in the diving animal.

81

Daly & Scott (1958, 1963, 1964) and Daly & Hazzledine (1963) showed that these primary responses were usually masked in freely breathing animals by secondary effects arising from simultaneous respiratory stimulation. Thus, under conditions in which stimulation of the carotid and aortic bodies occurs in response to cessation of breathing, that is in apnoeic asphyxia, the progressively increasing hypoxia and hypercapnia produced the primary responses of brady-cardia and peripheral vasoconstriction, which were shown to have originated reflexly from chemoreceptor stimulation (Angell-James & Daly, 1969a).

Considerable species difference exists in the response to carotid body stimulation in spontaneously breathing animals. In general terms, a relatively great increase in pulmonary ventilation results in either little change in heart rate or a slight tachycardia and peripheral vasodilation. In those species in which respiration is not so much increased, bradycardia is usually seen. Thus, dogs and monkeys show slight tachycardia and hyperventilation, while cats and har-bour seals experience bradycardia and smaller increases in ventilation (Fig. 5.2).

Daly and colleagues have recently carried out a series of studies of cardiorespiratory controls in diving harbour seals. Experimental dives were produced by flooding the face of the animal through a hollow plastic mask. Diving responses were elicited when the level of anaesthesia was adjusted to permit spontaneous respiration. The carotid sinus and carotid body regions were isolated from the circulation in a manner similar to that employed in earlier dog experiments with provision for perfusion of the carotid body with blood from either the arterial circulation directly or with blood which had cir-culated through an extracorporeal gas exchange device, thus permitting experimental alterations in the gas composition of the perfusing blood.

Face immersion by flooding the mask produced an im-mediate diving response which consisted of apnoea in the expiratory position and bradycardia with an accompanying slight fall in arterial blood pressure. As asphyxia progress-ively increased, the chemoreceptor drive was experimentally withdrawn by changing the perfusion of the carotid body

82

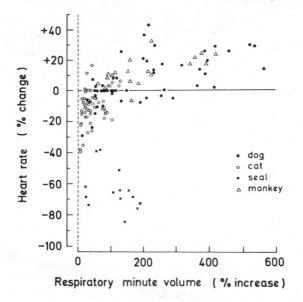

Fig. 5.2. Change in heart rate and respiration produced by stimulation of carotid bodies by perfusion with hypoxic blood (MacLeod & Scott, 1964; M. de B. Daly, J. E. Angell-James & R. Elsner, unpublished results; M. de B. Daly, personal communication).

from the seal's hypoxic, hypercapnic blood to blood of high P_{O_2} and normal P_{CO_2} from the external oxygenator. The heart rate was immediately restored to its pre-dive value. Reperfusion with the seal's own blood without removal of the water from the animal's face re-established the diving condition of bradycardia but did not stimulate lung ventilation (Daly *et al.*, 1977) (Fig. 5.3). Whatever influence there may be on these outputs by aortic body chemoreceptor stimulation in seals is unknown. Vascular reflex effects resulting from aortic body stimulation in dogs were similar to those of the carotid bodies, but respiratory responses were weaker (Daly & Ungar, 1966).

The experiments demonstrated that (1) initial apnoea and bradycardia were produced by water stimulation, most likely acting through facial trigeminal receptors, (2) carotid body stimulation sustained the diving bradycardia during the ensuing progressive asphyxia and (3) the carotid chemo-

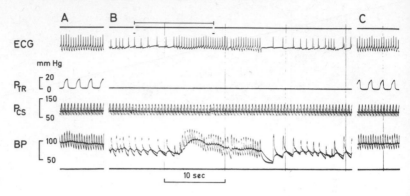

Fig. 5.3. Withdrawal of harbour seal *Phoca vitulina* carotid body chemo-receptor drive during apnoeic asphyxia. A, control. B, arterial blood P_{O_2}, 30 mmHg; P_{CO_2}, 88 mmHg; pH 7.07. Signal mark, perfusion of carotid bodies with oxygenated blood. C, control 4.5 min after recommencing artificial respiration. ECG, electrocardiogram; P_{TR}, intratracheal pressure; P_{CS}, carotid sinus perfusion pressure; BP, Phasic and mean arterial blood pressure. (From Daly, Elsner & Angell-James, 1977.)

receptor respiratory drive was inhibited during the dive, probably by central action originating from trigeminal exci-tation. This last conclusion is discussed more fully under 'Reflex interactions', below.

A series of studies has recently shown that the differences between the onset of the diving responses in ducks and seals can be explained by the relative contributions from stimuli to the nasopharyngeal–glottal region and the chemoreceptor inputs. Those differences lie principally in the rate of onset of the cardiovascular changes upon immersion, the initial response being slower in ducks and requiring at least 1 min for full development. Butler & Taylor (1973) and Bamford & Jones (1974) have shown that water is not a necessary accompaniment to the development of the full diving re-sponse in ducks. When apnoea was experimentally produced in curarized, relaxed ducks without the presence of water, the subsequent cardiovascular responses were not significantly different from those produced by water immersion. The fundamental property of water in the elicitation of the responses apparently lies simply in the reflex production of apnoea. (Reviews: Jones, 1976; Jones & West, 1978).

84

Blix, Lundgren & Folkow (1975) showed that the typical initial slight bradycardia in diving ducks is accompanied by just sufficient peripheral vasoconstriction that arterial blood pressure is well maintained. Blix (1975) also showed that this initial cardiovascular response is independent of asphyxia. Cardiac ventricular receptors may play a role in the initiation of bradycardia in diving ducks (Blix, Wennergren & Folkow, 1976b), but Jones, Milsom & West (1980) failed to confirm that possibility.

Lung inflation and related reflexes

Several lines of evidence show that the full development of cardiovascular responses to diving is critically dependent upon a state of pulmonary rest or absence of deformity in the lungs. Both diving and terrestrial species have been shown to respond to lung inflation by increasing heart rate, the presumed result of stimulating pulmonary stretch receptors: dogs (Anrep *et al.*, 1936a; Daly & Scott, 1958), ducks (Andersen, 1963b), geese (Cohn, Krog & Shannon, 1968) and seals (Angell-James, Elsner & Daly, 1981). A vasodilator response was also shown to occur in dogs (Daly & Robinson, 1968). Spontaneous breathing and sudden inflation of the lungs with normal tidal or larger volumes appear equally effective in producing the response. Quiet, shallow breathing, hypoventilation and low volume artificial respiration result in little or no stimulation of the reflex circuitry. The effect in ducks was less when the lungs were inflated with asphyxic gas mixtures (Andersen, 1963b).

During apnoea, whether from simple breath-holding or from submersion, the reduced afferent input from pulmonary receptors along with the associated cessation of central inspiratory neuronal activity permits the full unrestrained expression of chemoreceptor vagal effects in a variety of terrestrial mammals (Anrep *et al.*, 1936a; Daly, Hazzledine & Ungar, 1967; Angell-James & Daly, 1973; Kordy, Neil & Palmer, 1975; Daly *et al.*, 1978). Similarly, lung inflation has the effect of suppressing cardiovascular influences that result from stimulation of trigeminal receptors (Angell-James & Daly, 1978; Gandevia, McCloskey & Potter, 1978). Lopes &

Palmer (1976) proposed a central 'gating' hypothesis to explain the control of neural signals that originate from pulmonary vagal afferent impulses and result in increased heart rate with lung inflation.

Inflation of the harbour seal's lungs with volumes equal to or exceeding normal tidal volumes during the expiratory phase of spontaneous breathing produced immediate increases in heart rate. Deflation reversed the process. Similar cardiac responses resulted from inflation and deflation of the lungs during experimental dives. Furthermore, these cyclic changes in heart rate could be produced by inflation and deflation when apnoea and bradycardia were the result of combined stimulation of carotid chemoreceptors and trigeminal or laryngeal nerves (Angell-James *et al.*, 1981). It would appear from these lines of evidence that the decrease or cessation of pulmonary afferent activity and the associated decline in central inspiratory drive are potentially important contributors to the neural integration of the diving response. Lung inflation responses, like those of chemoreceptor, trigeminal and laryngeal origin, are qualitatively similar in terrestrial and marine mammals and birds, but there tends to be a greater response in aquatic species.

Arterial baroreceptor reflexes

The possible role of arterial baroreceptor reflexes in the development and maintenance of cardiovascular responses to diving has been a topic of considerable interest and the subject of several experimental investigations. Studies in diving seals show that bradycardia and a related decline in cardiac output take place with a simultaneous increase in peripheral resistance and little overall change in mean arterial pressure (Irving *et al.*, 1942*a*; Elsner *et al.*, 1966*a*). Ducks respond similarly (Johansen & Aakhus, 1963). On the basis of these and other observations, it was generally assumed that the baroreceptor reflexes have little to do with the initiation and control of diving responses. Furthermore, diving bradycardia was found to persist in ducks with denervated baroreceptors (Jones, 1973). Sympathetic blockade of the peripheral vasoconstrictor response in submerged ducks was ineffective in eliminating bradycardia (Kobinger & Oda, 1969; Andersen & Blix, 1974; Blix, Gautvik & Refsum, 1974). In fact, the marked decrease in cardiac output might have been expected to produce activation of baroreceptor reflexes resulting in tachycardia rather than bradycardia.

The view that baroreceptor reflexes might not be involved was, however, a deceptively superficial one, and the elucidation of their possible role began with the suggestion by Angell-James & Daly (1972b) that they are reset to a different level of response during diving. By this was meant that, as a closed-loop system, the plot which relates mean arterial pressure and heart rate would be changed in such a way that heart rate at the same reference pressure would be slower in the diving condition. They also suggested that the possibility of an associated change in sensitivity or gain of the reflex should be looked for.

Experiments with harbour seals revealed that carotid sinus baroreceptors and their reflex responses in that species are essentially the same as those seen in terrestrial mammals. Specifically, the responses to both pulsatile and non-pulsatile perfusion of the carotid sinus region resulted in baroreceptor impulse discharge characteristics similar to those earlier reported to exist in the other mammals (Angell-James, Daly & Elsner, 1978) Also, the relations between changes in heart rate and arterial pressure were found to be similar to those described in dogs. The maximum sensitivity of the derived reflex control of heart rate (Δ mean arterial pressure/Δ mean carotid sinus pressure) was 1.2 in seals compared with 1.2 (Schmidt, Kumada & Sagawa, 1972) and 2.0 (Scher & Young, 1963) in dogs.

The study demonstrated that, as predicted, the baroreflex was reset in the direction of bradycardia, that is, a lower heart rate at the same reference arterial pressure. In addition, stimulation of baroreflexes by administration of phenylephrine in doses resulting in discrete, brief pressure elevations, and stimulation of the carotid sinus baroreceptor reflex by elevating the pressure in the vascularly isolated sinus both resulted in an increased pulse interval. In some of the tests a linear relation was found to exist between systolic or mean arterial pressure and pulse interval, while other experiments

87

showed a curvilinear relation with a steepening slope at higher pressures. The gain determined at the value of reference systolic pressure was about 3.6 ms/mmHg (Δ pulse interval/Δ systolic arterial pressure). This value compares with 46.5 ms/mmHg during the experimental dives. Similar changes in gain were noted when calculated for mean arterial pressure.

The results were similar whether the reflexes were examined as open or closed loops. The changes described, the resetting and changed gain of baroreflexes, might be accounted for by a shifting to the left of the relation represented in the S-shaped curves relating heart rate and blood pressure. That would mean that what is described as increased gain would actually be a further resetting of the reflex. Whatever description is applied to the phenomenon, it appears that the degree of heart rate slowing in diving animals is determined in part by the arterial baroreceptor reflexes. The extent to which they operate will depend in turn on the arterial pressure achieved by peripheral vasoconstriction which is in effect while the cardiac output is reduced, and by shifts in the reflex control of heart rate brought about by other inputs.

Reflex interactions

In recent years it has become appreciated that the reflex responses that occur as a result of excitation of a homogeneous group of receptors can be considerably modified if the same receptors are stimulated while a second sensory input is simultaneously introduced into the nervous system. The interaction between two or more simultaneously occurring reflexes produces complex responses, and their analysis can prove difficult. For example, reactions to combined stimulation are often either greater or less than the simple algebraic sum of their separate effects. Some of the more important interactions associated with breath-hold diving which are involved with regulation of the cardiovascular system are elicited by simultaneous inputs from trigeminal receptors, from peripheral arterial chemoreceptors and from inputs arising from concomitant changes in a lung ventilation.

An essential link between the results of experiments on

terrestrial species and diving animals is related to the reflexes engendered by respiratory activity (Daly, 1972; Angell-James & Daly, 1972a; Daly & Angell-James, 1975). Examination of the reflex origins of sinus arrhythmia has shed some light on the interactions between the respiratory centre and the neurons governing the circulation: the cardio-inhibitory centre and the vasomotor centre. During normal spontaneous inspiration respiratory centre activation exerts a blocking influence on the cardio-inhibitory centre resulting in tachycardia (Anrep *et al.*, 1936b). This effect has been experimentally examined and verified by Joels & Samueloff (1956). It is reinforced during inspiration by stimulation of pulmonary stretch receptors (Anrep *et al.*, 1936a).

During expiration the cessation of the central inspiratory drive and the reduced discharge from pulmonary inflation receptors results in greater activity of the cardio-inhibitory centre, and bradycardia ensues (Anrep *et al.*, 1936a). Trigeminal excitation of expiratory neurons in the respiratory centre also directly stimulates the cardio-inhibitory centre, further depressing the heart rate (Cordier & Heymans, 1935). Thus, the cardio-inhibitory centre is under the potential influence of three separate reflex respiratory neural inputs: a lung stretch reflex, whose influence is diminished during expiration or apnoea in the expiratory position, respiratory centre direct effects and reflexes originating in the trigeminal (and possibly other) receptors. Furthermore, carotid body stimulation during expiration enhances the activity of the cardio-inhibitory centre (Haymet & McCloskey, 1975). Vasomotor centre activity is probably also under a regulating influence from these reflexes. The combined effects would produce the cardiovascular reactions of the diving response by the combination of a cessation of rhythmic breathing and stimulation of the nasopharynx.

The parallel between the studies of diving species and the experimental work of Angell-James & Daly on dogs is clearly evident. They had demonstrated that cardiovascular responses resulting from apnoeic asphyxia were reflexly produced by stimulation of peripheral arterial chemoreceptors. Dogs were also shown to be responsive to stimulation of the superior laryngeal nerve, the combined stimuli producing a

Control mechanisms: diving ducks and seals

potentiated response (Angell-James & Daly, 1969*a*, 1972*a*, 1975).

One major problem has concerned the resolution of apparently contradictory influences produced by stimulation of the peripheral arterial chemoreceptors. This arises from the fact that hypoxic, hypercapnic stimulation of the carotid chemoreceptors produces a vigorous respiratory drive in the spontaneously breathing animal. It has been shown by Elsner *et al.* (1970*b*) and by Kerem & Elsner (1973*b*) that experimentally dived Weddell and harbour seals experience progressively increasing asphyxia in which the arterial P_{O_2} and P_{CO_2} reach levels (10 mmHg and 100 mmHg respectively) which are capable of producing a vigorous ventilatory drive in a non-diving seal (Bainton, Elsner & Matthews, 1973; Påsche, 1976). Earlier, Robin *et al.* (1963) showed that the ventilatory response of seals to increased alveolar carbon dioxide, though less than that in man, is still vigorous at high alveolar P_{CO_2} values. Correction for body weight reduces but does not eliminate the difference. The persisting ventilatory drive must necessarily be inhibited during diving.

The possibility that suppression of the chemoreceptor respiratory stimulating reflex during diving was due to an inhibitory input from the trigeminal receptors was first shown by experiments on dogs (Angell-James & Daly, 1973). Similar results have been obtained from diving seals. Daly *et al.* (1977) showed that during diving the increasing chemoreceptor drive did not affect the apnoea engendered reflexly by the trigeminal input. In other experiments in which the carotid bodies were stimulated by small injections of cyanide, the resulting hyperventilation was abolished by the trigeminal stimulation of an experimental dive. At the same time the chemoreceptor-induced bradycardia was greatly potentiated (Elsner *et al.*, 1977; Daly, Angell-James & Elsner, 1980). These results show, therefore, a differential action of the trigeminal input on the carotid body respiratory stimulating reflex and the carotid body cardiac vagal mechanism (Fig. 5.4).

The counterparts of these experiments are those carried out in terrestrial mammals in which it has been shown that chemoreceptors are less effective in stimulating breathing when the stimuli are delivered during the expiratory phase of

90

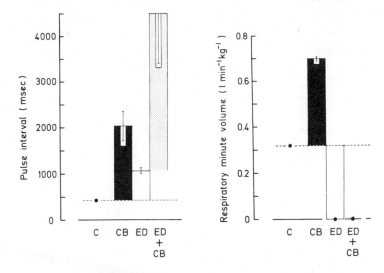

Fig. 5.4. Stimulation of harbour seal *Phoca vitulina* carotid body chemo-receptors with cyanide, CB, experimental dives, ED, and combined stimulations, ED + CB; controls, C. Effects on pulse interval and respiration. Mean values ± S.E. are shown for ten experiments with five seals. (From Elsner, Angell-James & Daly, 1977*a*.)

breathing than during inspiration (Black & Torrance, 1971; Eldridge, 1972; Haymet & McCloskey, 1975). Moreover, the cardio-vagal mechanism that results from chemoreceptor stimulation is inhibited by the central inspiratory drive (Haymet & McCloskey, 1975) and by stimulation of the pulmonary stretch receptors through phasic changes in the lung volume (Daly & Hazzledine, 1963; Angell-James & Daly, 1969*a*, 1978; Haymet & McCloskey, 1975). In other words, apnoea reflexly induced in the expiratory position by the trigeminal input leaves the cardiac vagal mechanism unmodulated through a combination of cessation of the central respiratory drive and diminution in the pulmonary stretch receptor activity (Angell-James & Daly, 1969*a*; Haymet & McCloskey, 1975).

Higher cerebral influences

Earlier references to the considerable variability in diving responses, depending upon whether the dives are forced,

91

trained or voluntary, suggest that conscious awareness, fear and preparation may contribute to the programming of the physiological resource and reserve which is appropriate for a given diving situation. Consider, for instance, the Weddell seal, capable of submerging for more than 60 min, on a prolonged exploratory under-ice foray. Not only must it perform prodigious feats of navigation if it is to return to its breathing hole, but it must also sense the half-time of its oxygen reserves in order to determine when to head back!

Discrete stimulations in the hypothalamus of elephant seals (Van Citters *et al.*, 1965) resulted in respiratory and cardiovascular responses similar to those seen during dives in ducks (Folkow & Rubinstein, 1965). While the results do not positively identify the central structure responsible for controlling these reactions, they suggest an avenue of investigation which should be promising in a search for the central mechanisms involved in the diving response.

Diving control mechanisms are summarized in Fig. 5.5.

HUMAN STUDIES

Face immersion

The importance of face immersion in the development of the diving response in man was initially suggested by two observations. Simple face immersion with breath-holding was found to be as effective as total body immersion in producing diving bradycardia, and the combination of face immersion plus breath-holding produced a greater bradycardia than breath-holding alone (Elsner *et al.*, 1963; Irving, 1963). Both of these conditions have been confirmed by others as providing the adequate stimulus (Wolf *et al.*, 1965; Kawakami *et al.*, 1967; Whayne & Killip, 1967; Corriol & Rohner, 1968*a*; Heistad *et al.*, 1968; Campbell *et al.*, 1969*a*; Song *et al.*, 1969; Moore *et al.*, 1972*a*). However, some suggest that there is no difference between the responses to breath-holding alone and breath-holding with face immersion (Brick, 1966; Paulev, 1968*a*; Burch & Giles, 1970; Folinsbee, 1974).

It has also been shown repeatedly that face immersion

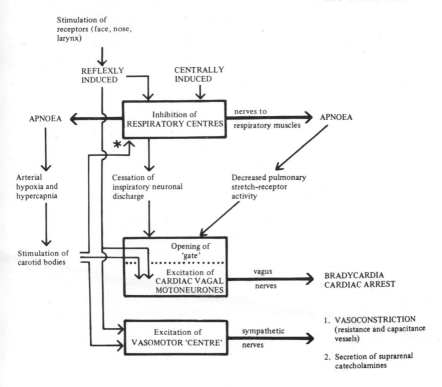

Fig. 5.5. Summary of control mechanisms operating during diving. Afferent inputs and sites of possible interactions are shown. The same basic scheme is applicable to birds and mammals, including man. Baroreceptor inputs are not indicated. (From Daly & Angell-James, 1979.)

without breath-holding evokes bradycardia, although the response is usually smaller and transient compared with that produced by face immersion with breath-holding (Sasamoto, 1965; Brick, 1966; Kawakami *et al.*, 1967; Corriol & Rohner, 1968*b*; Paulev, 1968*a*). Brick (1966) showed that this response was not caused by a change in breathing pattern. Furthermore, if the face is removed from the water but the breath is still held, the heart rate and forearm blood flow responses recover towards the pre-procedure levels (Gooden, Lehman & Pym, 1970; Strφmme *et al.*, 1970) Fig. 5.6.

Stimulation of the skin in the region of the eyes and nose appears to be particularly important. Strφmme *et al.* (1970)

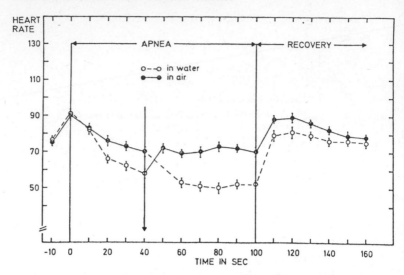

Fig. 5.6. Heart rate responses to human breath-holding alternately in air and in water. Transition from air to water or from water to air indicated by arrow. (Heart rate in beats/min.) (From Strømme, Kerem & Elsner, 1970.)

examined face immersion breath-holding with and without a face mask. Initially the subjects wore a face mask that covered both eyes and nose for 40 s, then removed the mask for the last 60 s of the procedure (Fig. 5.6). A moderate bradycardia was established and plateaued during the first 40 s, but upon removal of the face mask the heart rate decreased further. Scuba divers in whom water had been in contact for some time with the skin of the face, except for the region around the eyes and the nose, still developed significant bradycardia upon removal of the face mask while breath-holding continued (Gooden *et al.*, 1975). Neither breath-holding alone nor mask removal with continued breathing produced as marked a bradycardia. Wearing a face mask and swimming cap attenuated the usual heart rate and forearm blood flow responses to total immersion with breath-holding (Gooden *et al.*, 1970). The area of skin covered is innervated by the ophthalmic division of the trigeminal nerve. Russetzki (1925) observed bradycardia in some subjects when an electrical current was applied to branches of the trigeminal nerve.

94

In an attempt to define more accurately the relative importance of different areas of afferent input from the face, Parfrey & Sheehan (1975) prevented water touching selected areas by covering them with moulded acrylic devices. When nose, eyes and mouth were simultaneously covered, the reduction in heart rate on immersion was not significantly different from that produced by breath-holding only. Irrigation of the nasal passages with cold water or saline produced apnoea and a rise in blood pressure (Ebbecke, 1944). However, the diving response could still be elicited in human subjects even though they were wearing a nose clip which prevented water from entering the nostrils (Campbell *et al.*, 1969*a*).

There appears to be some adaptation of the sensory input with increasing duration of face immersion, and that might explain the transitory nature of bradycardia evoked by face immersion without breath-holding (Brick, 1966; Kobayasi & Ogawa, 1973). Similarly, $2\frac{1}{2}$ to 3 min of face immersion with snorkel breathing preceding a period of breath-holding has been shown to attenuate the heart rate and forearm blood flow responses during the breath-hold (Whayne & Killip, 1967; Campbell *et al.*, 1969*a*).

The nature of the sensory receptors involved is still not clear, but temperature sensation appears to play an important role. Craig (1963) noted that the bradycardia response to breath-hold diving in water at 22 °C was greater than at 29.5 and 34.5 °C. Korean Ama divers developed more pronounced bradycardia during diving in winter at a water temperature of 10 °C than during diving in summer at 27 °C (Hong *et al.*, 1967). Several workers have confirmed the potentiation of bradycardia resulting from face immersion with breath-holding at lower water temperatures (Kawakami *et al.*, 1967; Whayne & Killip, 1967; Paulev, 1968*a*) especially those below 20 °C (Song *et al.*, 1969; Moore *et al.*, 1972*a*; Parfrey & Sheehan, 1975).

Significant bradycardia and reduction in forearm blood flow occur during total immersion with breath-holding in water at 34 °C (Campbell *et al.*, 1969*b*). Water at this temperature feels neither warm nor cool, suggesting that sensory modalities other than temperature are involved.

95

Separating the face from water by a thin plastic sheet attenuated the bradycardia response (Wolf *et al.*, 1965; Moore *et al.*, 1972*a*; Hunt, Whitaker & Willmott, 1975; Parfrey & Sheehan, 1975; Gooden, Holdstock & Hampton, 1978) even when skin and water temperatures are equal. Facial wetting, therefore, appears to be a stimulus for initiating the diving response itself.

Cessation of breathing

The usual cardiovascular changes that occur with diving depend upon the concurrence of face immersion and breath-holding (Campbell *et al.*, 1969*a*). In human diving the subjects usually voluntarily stop breathing just before the face enters the water as a result of conscious inhibition of respiration by the higher centres of the brain. Reflex inhibition of breathing may also occur in response to water touching the face. Hamilton & Mayo (1944) made the following observation in a study of vital capacity during immersion.

> These experiments were performed with a nose clip firmly in place and a mouthpiece connected to a spirometer by a long tube. When the subject had settled underwater with the understanding that he was to inhale as soon as he had exhaled completely he found that it was nearly impossible to make the effort. This was not because he was so deep underwater that he could not accomplish the muscular act of expanding the chest but was due to stimulation by the water on the face or some similar factor that prevented him even trying to inhale . . . The force of the inhibition in man is not very strong because it may be broken through by a firm effort of the will.

Breath-holding involves several physiological events which probably relate to the initiation of the diving response in man: intrathoracic pressure and its effect on venous return, blood gas changes, the absence of rhythmical stimulation of the pulmonary stretch receptors and the direct effect of inhibition of the respiratory centre on the cardiovascular neurons in the brain stem.

If, after the development of the diving response, the subject resumes breathing via a snorkel, but face immersion continues, there is still a rapid increase in heart rate and forearm blood flow from the levels during diving (Gooden, 1971*b*). However, relief of asphyxial blood gas changes alone is not sufficient for the recovery in heart rate following a dive. Daly & Angell-James (1975) found that the heart rate still increased rapidly upon resumption of breathing after face immersion with breath-holding even though they maintained the pulmonary gas composition at the breaking-point level by requesting the subject to breathe a gas mixture containing 10.5% oxygen and 6.5% carbon dioxide. This leaves rhythmical stretching of lung receptors and the release of central inhibition of the respiratory centre as likely mechanisms in recovery of the heart rate.

Depth of dive, intrathoracic pressure, venous return

Wyss (1956*a*) suggested that heart slowing was more pronounced at greater underwater depths, but subsequent work has failed to substantiate this claim (Craig, 1963; Wolf *et al.*, 1965; Moore *et al.*, 1972*a*). Kerem & Salzano (1974) studied the effects of breath-holding only and face immersion with breath-holding at ambient air pressures up to 27 atm. No direct correlation was found between the ambient air pressure and the bradycardial response to either procedure. Hong *et al.* (1973) believed that they had found a potentiation of the bradycardial response to these procedures in subjects after breathing a hyperbaric helium–oxygen mixture. They suggested that the facial receptors may have responded differently to a given thermal stimulus in hyperbaric heliox environments.

Craig (1963) showed that tachycardia was produced by breath-holding at positive intrathoracic pressures up to 40 mmHg. The increase in heart rate was proportional to the increase in intrathoracic pressure. A small bradycardia was found by later workers during breath-holding at low positive or negative pressure (Paulev, 1968*a*; Song *et al.*, 1969). Face immersion with breath-holding produced bradycardia irrespective of oesophageal pressures ranging from -5 to $+16$ cm

97

H_2O (Kawakami *et al.*, 1967). Song *et al.* (1969) found that the heart rate was substantially lower during face immersion than during breath-holding in air over a range of intrapleural pressures from 0 to 40 cm H_2O.

Tachycardia produced by breath-holding at high intra-thoracic pressure can be converted into bradycardia by immersing a standing subject to the neck, and facilitation of venous return with consequent increased stimulation of arterial baroreceptors has been suggested as the explanation (Craig, 1963; Harding *et al.*, 1965). The implication is that the effects of raised intrathoracic pressure and immersion to the neck have opposite effects upon venous return. Song *et al.* (1969) found progressively greater bradycardia during breath-holding in the standing subject as water depth was increased to just below the nose. When the whole head was submerged the heart slowed even further. However, Moore *et al.* (1972*a*) showed that immersion of only the face with breath-holding produces bradycardia equal to that produced by total immersion in the standing subject. It therefore appears that facilitation of venous return can be eliminated as a factor in the development of diving bradycardia.

Lung volume

Elsner *et al.* (1963) reported that the cardiovascular response produced by face immersion with breath-holding was accentuated when the procedure was performed in the expiratory position. A similar observation was made by Asmussen & Kristiansson (1968). However, Kawakami *et al.* (1967) saw no difference in the magnitude of bradycardia during face immersion with breath-holding at the end of expiration or full inspiration.

Baroreceptors

Several authors have suggested that diving bradycardia in man is secondary to the rise in arterial blood pressure (Craig, 1963; Harding *et al.*, 1965; Paulev, 1968*a*). However Heistad *et al.* (1968) found that even marked reductions in heart rate were observed on occasions without a rise in arterial blood pressure. If diving bradycardia is secondary to the increase in arterial blood pressure, a correlation between the fall in heart rate and the rise in arterial pressure during diving might be expected. Song *et al.* (1969) and Gooden (1972) found no significant correlation between the changes in mean arterial blood pressure and heart rate during total immersion. However, human subjects may respond like seals (Angell-James *et al.*, 1978) in that the baroreflex relating heart rate and mean arterial blood pressure at a given reference pressure is reset towards bradycardia and the sensitivity increased. Baroreceptor resetting and a possible increase in sensitivity would explain the observation of Heistad *et al.* (1968) that marked diving bradycardia could occur without a rise in arterial blood pressure. Usually, however, mean arterial pressure increases in man during dives (Campbell *et al.*, 1969*a*, *b*).

Chemoreceptors

Olsen *et al.* (1962*a*) pointed out that diving bradycardia in man occurred before any major changes in arterial oxygen or carbon dioxide tensions could be expected and that in general the degree of bradycardia did not correlate with the duration of apnoea. The asphyxial blood gas changes produced during total immersion were no greater than those produced during breath-holding with the head out of water (Olsen *et al.*, 1962*b*). Hypocapnia, produced by vigorous hyperventilation of air before total immersion, did not prevent development of bradycardia or increased arterial blood pressure during the subsequent dive (Olsen *et al.*, 1962*b*). Heart rate data obtained by Kawakami *et al.* (1967) during breath-holding with face immersion after two minutes' inhalation of a hypercapnic gas mixture suggest that the bradycardial response was no different from that produced after breathing air. During simulated diving in subjects with arterial hyperoxia, the heart rate and forearm blood flow fell during the first minute but then recovered towards the control level despite progressive arterial hypercapnia (Elsner *et al.*, 1971*a*). Thus arterial carbon dioxide levels do not appear to be of any great importance in the development of the diving response in man.

Several reports suggest a potentiation of diving bradycardia when dynamic exercise is performed immediately before and during face immersion with breath-holding (Strømme *et al.*, 1970; Bergman, Campbell & Wildenthal, 1972). In addition it has been demonstrated that breath-holding immediately before face immersion potentiates the bradycardia response to apnoeic face immersion (Strømme *et al.*, 1970; Gooden *et al.*, 1970) Fig. 5.6. Very profound reductions in heart rate have been produced by combining these two factors (Strømme *et al.*, 1970; Strømme and Blix, 1976). Also, a modest attenuation of the heart rate and forearm blood flow responses to face immersion with breath-holding resulted when subjects breathed normally on pure oxygen for 15 min, thus raising their arterial oxygen tension to 352 ± 72 mmHg before the procedure (Elsner, Gooden & Robinson, 1971*b*). Similarly, Moore *et al.* (1972*b*) reported a significant reduction of the bradycardial response to apnoeic face immersion when the alveolar tension of oxygen was in a range from 300 to 450 mmHg.

Moore *et al.* (1972*b*) performed a series of experiments in which they selectively altered the alveolar partial pressures of oxygen and carbon dioxide by air or oxygen pre-breathing, exercise and rebreathing manoeuvres. During both rest and exercise, the bradycardia response was potentiated in the presence of low alveolar oxygen tension, and appeared to be independent of alveolar carbon dioxide tension. When the heart rate responses obtained in these studies are plotted against arterial or alveolar P_{O_2}, the line of best fit bears a resemblance to curves relating to the neural discharge from the carotid body and the alveolar P_{O_2} (Hornbein, 1968; Biscoe, Purves & Sampson, 1970). Exercise with superimposed face immersion and breath-holding leads to a greater degree of alveolar hypoxia and of bradycardia than the same procedure during rest. These findings provide an explanation of the potentiation of the bradycardia when exercise and breath-holding precede face immersion with breath-holding.

The evidence to date therefore suggests that the arterial oxygen tension acting via the peripheral chemoreceptors can modify the intensity of the diving response. The human experiments are thus reminiscent of studies of diving seals which are similarly responsive to alterations in blood gas composition, as demonstrated by reversal of diving bradycardia when the isolated carotid body is perfused with blood containing normal gas tensions (Daly *et al.*, 1977; Elsner *et al.*, 1977*a*).

Cardiovascular controls

Diving bradycardia in man can be blocked by atropine, indicating that the parasympathetic nerve supply plays a critical role in the development of this chronotropic response (Heistad *et al.*, 1968). Frey & Kenney (1974) studied systolic time intervals in a group of young experienced swimmers during face immersion with breath-holding. Their results support the idea that diving bradycardia in man does not require the withdrawal of sympathetic tone as judged by the pre-ejection period. Experiments of Finley, Bonet & Waxman (1979) indicated that hypertension was not essential for the development of bradycardia during human face immersion and it was not altered by alpha or beta sympathetic blockade.

The reduction in finger, forearm and calf blood flow observed during face immersion with breath-holding and total body immersion has been assumed to result from an increase in sympathetic adrenergic nerve activity to the arterial vasculature in the limbs. Heistad *et al.* (1968) found that the vasoconstrictor response was more pronounced in the finger than the forearm. They considered the possibility that such a difference might result from the activity of sympathetic cholinergic vasodilator nerves to the skeletal muscle. Such a response might be evoked by emotional stress and oppose the vasoconstriction (Blair *et al.*, 1959), but they did not find any augmentation of forearm vasoconstriction in subjects to whom atropine had been administered. Another possible explanation is that vasoconstriction may be more intense in the skin than in the skeletal muscle vasculature; finger flow has a much higher proportion of skin flow than does forearm flow.

101

Higher centres and general anaesthesia

The diving response in man is particularly susceptible to the influence of the higher brain centres. Fear appears to potentiate the response (Wolf *et al.*, 1965). Wolf *et al.* (1965) reported that bradycardia could occur immediately after the order to immerse the face was given to the subject and before the face had reached the water. The inference was that diving bradycardia in man could develop in anticipation of immersion as a result of higher centre activity. When the subjects were distracted or harassed, diving bradycardia often failed to occur despite face immersion. Although it has not been specifically tested, there is no reason to believe that conditioning of heart rate to the experimental situation could not take place.

It is well known that general anaesthesia blocks the diving response in animals, but little is known of its effects in man. Whayne *et al.* (1971) studied the effects of halothane anaesthesia on the cardiovascular response to facial application of a wet cloth at 0°C with simultaneous breath-holding. Anaesthesia decreased the bradycardial response from 29 to 3.6% below the control rate. They suggested that general anaesthesia abolished the parasympathetic component of this response and attenuated the sympathetic component.

Diving experience, physical fitness and age

Irving (1963) stated that diving bradycardia occurred with more regularity and to a greater degree in those accustomed to swimming and with a proclivity for aquatic activity. Several subsequent studies have supported this observation (Corriol, Rohner and Fondarai, 1966; Paulev, 1968*a*; Hong *et al.*, 1970; Kobayasi and Ogawa, 1973), while others have not (Craig, 1963; Harding *et al.*, 1965; Elsner *et al.*, 1966*a*). The magnitude of the diving response is reported not to be influenced by physical fitness (Whayne and Killip, 1967; Strømme *et al.*, 1970), but Bove *et al.* (1968) reported that a physical training programme lasting ten weeks augmented diving bradycardia.

Hong *et al.* (1970) found a significant correlation between the maximal bradycardia and the age of their subjects in face immersion experiments. The bradycardia became less pronounced with advancing age, and the aging effect was less marked in regular divers than in non-divers. Gooden *et al.* (1978) made similar measurements in patients with and without apparent heart disease, and incidentally found a correlation between the bradycardia produced by face immersion breath-holding and the subject's age.

6 PERINATAL ASPHYXIA AND SURVIVAL

> The problem of science will consist precisely in this, to seek
> the unitary character of physiological and pathological
> phenomena in the midst of the infinite variety of their
> particular manifestations (Claude Bernard, 1865).

The process of birth, involving the transition from placental
to lung respiration, exposes the mammalian infant to asphyxial
threats. Signs of its presence are revealed by measurement of
blood gases and pH in samples of blood obtained during birth
and in the observation of a transient or persisting decline in
heart rate. It is not surprising therefore that the terrestrial
animal fetus and neonate are more tolerant of asphyxia than
are adults of the same species. The interesting history of
knowledge concerning this advantage dates from experiments
on kittens by Robert Boyle in the seventeenth century (Dawes,
1968). Adaptations leading to oxygen sparing, an increased
dependence upon anaerobic metabolic processes and an
augmented capacity for tissue buffering of the accumulated
end-products have been identified. The major requirement
during oxygen deprivation, as with diving animals, is for
protection of the integrity of the least resistant vital organs,
the brain and heart. Some adaptive mechanisms appear
common to both perinatal animals and diving species. There
are, however, some important differences.

Fetal and neonatal resistance to asphyxia

While mammalian species vary greatly in their resistance to
asphyxia at birth, the general rule of superiority over adults of
the same species holds true. However, new-born seals, and
probably new-born marine mammals generally, tolerate
asphyxia less well than do adult animals. New-born elephant
seals *Mirounga angustirostris*, while exhibiting typical diving

104

bradycardia upon submersion, were clearly stressed by more than a few minutes' diving, which would have been easily endured by an adult (Hammond *et al.*, 1969). Diving ability is not well developed in infant harbour seals (Harrison and Tomlinson, 1960), fur seals (Irving *et al.*, 1963) or Weddell seals (Kooyman, 1968). Kooyman measured the duration of 104 dives by three free-swimming Weddell seal pups. The longest was 5 min, much shorter than the adult performance.

There are several possible explanations for the modest asphyxial resistance of infant seals. The pregnant Weddell seal was found to have a higher haemoglobin concentration, hence higher oxygen capacity, than its near-term fetus (Lenfant *et al.*, 1969*a*). This condition is the reverse of that in most terrestrial mammals, in which haemoglobin is more concentrated in the fetus (reviewed by Dawes, 1968). This relation in the Weddell seal is probably advantageous to the pregnant diving mammal because it would tend to reduce the requirement for uterine blood flow during submergence. As in most mammals, the oxygen affinity of fetal seal blood is higher than that of maternal blood. That is to say that the fetal oxygen–haemoglobin dissociation curve is shifted to the left. The P_{O_2} values at 50% saturation (P_{50}) of fetal and maternal blood were 22.1 mmHg and 28.5 mmHg in fetus and mother respectively. The blood oxygen capacity of the new-born seal steadily increases with early growth, thus augmenting oxygen storage.

Blood volume related to body weight is probably less in the new-born Weddell seal, since the large venous reservoir of the adult inferior vena cava is relatively undeveloped in that of the new-born seal (R. Elsner, unpublished observations). The bulbous enlargement at the root of the aorta, which plays a role as a circulatory *Windkessel*, is also poorly developed in infant Weddell seals and harbour seals (Rhode *et al.*, 1983).

The Weddell seal is born with a brain weighing only a little less than the adult brain (Table 6.1). This condition suggests that the fraction of the total available oxygen needed for brain metabolism during diving is greater in the new-born seal, thus increasing its minimum obligatory oxygen consumption. As growth proceeds, the brain to body weight ratio steadily declines, thus favouring the adult seal during long

105

Table 6.1. *Weddell seals: brain weight and body weight*

	Brain weight (kg)	Body weight (kg)	$\dfrac{\text{Brain wt}}{\text{Body wt}} \times 100$
Near-term fetus	0.41	32	1.3
New-born seal	0.40	29	1.4
New-born seal	0.39	29	1.3
Adult (4 yr)	0.56	405	0.14
Adult (5 yr)	0.60	418	0.14
Adult (5 yr)	0.50	395	0.13
Adult (6 yr)	0.53	400	0.13

dives. The same advantage seems not to apply or to be obscured by other conditions in the adult terrestrial mammal.

Adequate brain circulation is important for survival of the asphyxiated fetus and neonate (Dawes, 1968). Perinatal brain tissue can survive for some time in the absence of oxygen if adequate glucose is available as a substrate for anaerobic metabolic processes. Normally very little glucose is stored in brain tissue, but considerable amounts of glucose stored as cardiac muscle glycogen play an important role (Dawes, Mott & Shelley, 1959). The integrity of the circulation is essential to provide transport for glucose from the heart muscle to the brain. Survival is shortened by blocking glycolysis with iodoacetate and prolonged by the administration of additional glucose when accompanied by base to maintain near normal pH (Dawes *et al.*, 1963). Moreover, the brain of the new-born dog was found to have a specific resistance to total brain ischaemia, which enables it to survive longer than the similarly treated adult brain (Kabat, 1940). Birth asphyxia is apparently not limited to mammals, since the birth of a bird during hatching from the egg is accompanied by asphyxial changes in the gas composition within the egg (Tazawa, Mikami & Yoshimoto, 1971).

Diving seals and fetal lambs

Scholander (1960) suggested that fetal animals may react to asphyxia in a manner like that of diving seals. Barcroft (1946) recorded fetal bradycardia in response to asphyxia, and that observation has been reinforced by many investigators. Bradycardia has been evoked in the fetal sheep (Bauer, 1937) and human fetus by clamping the umbilical cord (Hon, 1966) or by compression of the fetal head (Hon, 1958). This bradycardia is believed to be reflex in origin with the vagus nerve providing the efferent limb (Mendez-Bauer *et al.*, 1963). Changes in the distribution of cardiac output similar to but generally less profound than those observed in seals have been demontrated in partially asphyxiated lambs delivered by caesarean section (Campbell *et al.*, 1967; Rudolph & Heymann, 1967; Dawes *et al.*, 1968) and in asphyxiated fetal monkeys (Behrman *et al.*, 1970). Similar compensatory changes in blood flow have been seen in studies of reduced umbilical circulation (Dawes & Mott, 1964; Brinkman, Mofid & Assali, 1974). Fetal cardiovascular adaptation to asphyxia and some similarities to adaptations of diving species are reviewed by Friedman and Kirkpatrick (1977). In some infants the peripheral vasoconstriction associated with these conditions may be of such intensity as to produce pathological ischaemia of the kidney (Dauber *et al.*, 1976) and gastrointestinal system (Fitzhardinge, 1977). Eventually, substrate depletion and acidosis occur and lead to an increased likelihood of brain damage. Infusion of alkali and glucose into asphyxiated new-born lambs and monkeys delayed the onset of that effect (Dawes *et al.*, 1963; Adamsons *et al.*, 1963).

A study of the blood flow in the abdominal aorta of a lamb during birth illustrates some of the cardiovascular responses to the transient asphyxia which accompanies that event. In this experiment the uterus of a near-term anaesthetized ewe (normal gestation lasts approximately 130 days) was surgically exposed and incised. A Doppler ultrasonic flow-transducer (Franklin *et al.*, 1963) was mounted on the fetal abdominal aorta. The incisions (fetal, uterine and maternal) were then closed around wire connections leading from the transducer to an externally mounted telemetry transmitter. Two weeks later normal birth of the healthy lamb occurred, and the blood flow in the fetal abdominal aorta was recorded throughout that episode (Fig. 6.1). Bradycardia and reduced blood flow can be seen during the transition to extrauterine

107

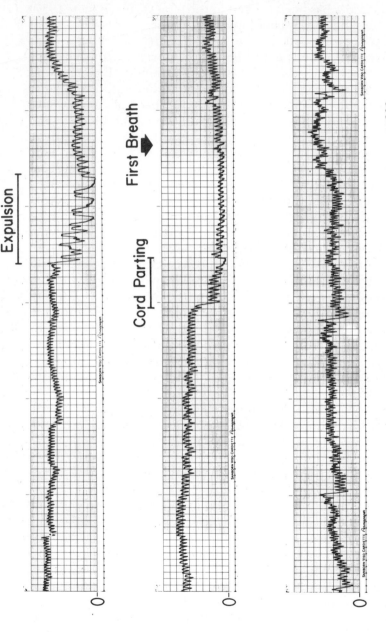

Fig. 6.1. Relative blood flow in the abdominal aorta of a lamb during birth. (From Elsner, 1978.)

life. Decreased blood flow was especially noticeable immediately after birth and before breathing was well established (Elsner, 1978).

The pregnant marine mammal

A prolonged dive by a pregnant marine mammal might be thought to represent an extreme example of both fetal and maternal adaptation to asphyxia. Some circulatory aspects of this adaptation in seals and its manifestations in terrestrial mammals have been investigated. Pregnant Weddell seals were compared with pregnant sheep with regard to maintenance of uterine blood flow during experimental dives and asphyxia. Comparative fetal responses were studied by determinations of selected blood flows and heart rates of fetal lambs and seals during maternal asphyxia (Elsner *et al.*, 1970*a*). Pregnant Weddell seals appear in considerable numbers along sea ice cracks near the shore of Ross Island, Antarctica, during late September and early October. The pupping begins a little later and reaches a peak about the end of October. Five seals in late pregnancy were studied in the wild by attaching to them recording depth gauges which were later recovered. Altogether 41 dives were recorded beneath the sea ice, each lasting longer than 20 min. The maximum dive duration was 60 min and the deepest was 310 m (Elsner, Kooyman & Drabek, 1969). These results compare well with those obtained by experiments on non-pregnant animals (Kooyman, 1966) and it is clear that a state of advanced pregnancy did not seriously interfere with diving performance.

In other experiments Doppler ultrasonic blood flow-transducers were implanted around the uterine arteries of four pregnant Weddell seals and on the renal artery of one animal. After recovery from surgery they were exposed to simulated dives lasting from 4 to 32 min. Renal flow was promptly reduced to about one-tenth or less of its normal level, as was typical of seals in earlier experiments (Elsner *et al.*, 1966*a*). However, uterine blood flow was reduced only slightly, if at all. Therefore, the diving pregnant seal maintained perfusion in the uterus, thus presumably

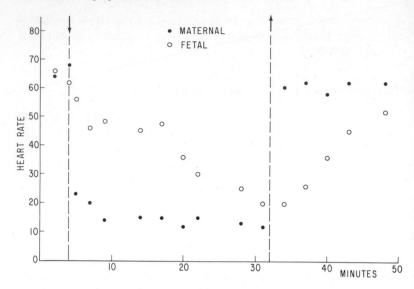

Fig. 6.2. Maternal and fetal heart rates (beats/min) during restrained dive of a pregnant Weddell seal *Leptonychotes weddelli*. Arrows indicate dive. (Elsner *et al.*, 1970*a*.)

sharing maternal oxygen resources with the fetus. Liggins *et al*. (1980) found similar blood flow distributions in forced dives of pregnant Weddell seals.

By appropriate placement of electrodes on the abdomen of the pregnant seal it was possible to obtain an electrocardiogram from which both maternal and fetal heart rates could be read. When this was done during a diving experiment, a striking difference between the two rates was evident. The maternal heart rate decreased precipitously and remained low throughout the experiment, as with diving seals generally. However, the onset of fetal bradycardia was much more gradual, the rate declining steadily as the episode continued (Fig. 6.2). The mechanisms governing the two responses appear to be quite different, those of the fetus probably being influenced by the levels of blood gases acting through chemoreceptors. The abrupt onset of maternal bradycardia argues for a more rapidly responding afferent neural signal (Elsner *et al.*, 1970*a*. However, Liggins *et al*. (1980) recorded fetal bradycardia before the onset of reduced fetal blood oxygen tension.

110

Some similarities to the reactions of seals were noted in asphyxial responses of pregnant and fetal sheep. The asphyxial experiment on pregnant seals was repeated on pregnant ewes, and blood flows changed in similar directions. Although the duration of asphyxia was only 1 min, necessarily much shorter than in the seal, uterine blood flow was well maintained, while renal flow was reduced to less than 10% of its control value (Fig. 6.3). Blood flow in the fetal abdominal aorta and the heart rate of the fetus gradually decreased after about 30 s (Elsner *et al.*, 1970*a*).

A basis for the maintained uterine artery blood flow might lie in local regulation of flow in the uterine arteries throughout pregnancy (reviewed by Comline & Silver, 1975). Resistance to flow in the uterus of pregnant sheep is not drastically changed by spinal anaesthesia (Huckabee, 1962; Ladner *et al.*, 1970) or by administration of the alpha adrenergic blocking agent, phenoxybenzamine, suggesting that sympathetic stimulation is relatively ineffective in producing vasoconstriction in that circuit (Greiss & Gobble, 1967).

Two separate mechanisms appear to be involved in the reactions of diving seals and asphyxiated fetuses. The rapid onset of the circulatory changes at the beginning of apnoea in seals is initiated by face immersion, probably acting through trigeminal sensory input (Daly *et al.*, 1977; Elsner *et al.*, 1977). A different mechanism is indicated for explanation of both the fetal responses to maternal asphyxia and the effects of umbilical cord clamping. A chemoreceptor explanation is likely, because the responses closely resemble those of the bradycardia and vasoconstriction arising from carotid body stimulation.

The analogy between diving seals and asphyxiated fetal or new-born sheep is tenable to the extent that the redistribution of cardiac output induced by interference with normal respiration represents a coordinated asphyxial defence mechanism. There are, however, clear differences in adaptations that favour survival, largely in the utilization of oxygen stores (Dawes, 1968). Blood oxygen is rapidly depleted during the first few minutes of fetal asphyxia. Thereafter, anaerobic glycolysis, depending in the main upon cardiac glycogen, provides energy for continued metabolic needs (Dawes *et al.*,

112

cm / sec

UTERINE

RENAL

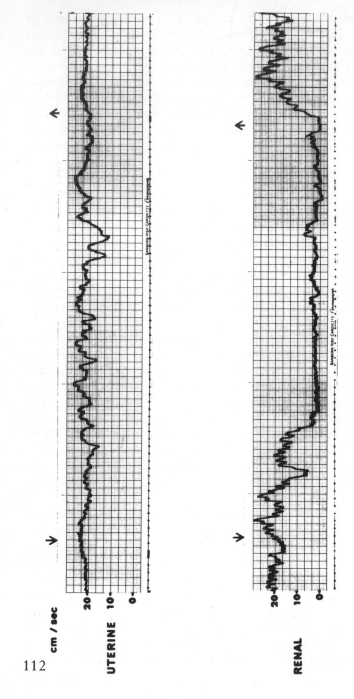

Fig. 6.3. Uterine and renal blood velocities in a pregnant sheep during experimental asphyxia lasting 1 min. Period of asphyxia indicated by arrows. (Elsner *et al.*, 1970*a*.)

1959). In this respect the fetus somewhat resembles the diving turtle for which anaerobic metabolism provides its major energy resource in long dives (Belkin, 1962). The cardiovascular responses are more profound in seals, and seals depend less on anaerobic mechanisms. Oxygen storage is clearly greater in seals, by virtue of both increased blood volume and haemoglobin concentration. Also the time courses of the observed events in the two species are widely disparate. Nevertheless, the fundamental circulatory mechanisms of asphyxial defence appear to be similar. Whether the blood transports oxygen or glucose to the brain, the prospect for survival is enhanced, as was originally pointed out by Le Gallois in 1812.

Chesler and Himwich (1944) showed that the relative anaerobic activity of various parts of the dog and cat brain varied with growth. The medulla of new-born animals had the highest rate, and it decreased with age while that of higher structures increased. Total brain metabolism was also less in infant than in adult animals. The brain stem of the seal's central nervous system may have relatively more resistance to asphyxia as its blood flow, measured by radioactive microsphere distribution during diving, is substantially less than that in the cerebral cortex (Elsner *et al.*, 1978).

Few experiments related to brain metabolism have been done in such a way as to allow direct comparisons among various species of new-born animals. However, studies of the electroencephalograms of Weddell seals during prolonged asphyxia might be pertinent. In both adult and infant animals that were asphyxiated or ventilated with nitrogen beyond the point at which the altered, but still reversible, EEG indicated the onset of impaired brain function (high-amplitude, slow-wave activity), the heart beat, and sometimes respiratory gasps, continued even though carotid arterial blood P_{O_2} had been reduced to less than 10 mmHg. Furthermore this endpoint was independent of arterial P_{CO_2} and pH over a wide range of values (Elsner *et al.*, 1970*b*). Thus, the indicated specific tolerance to low cerebral oxygen was similar in adult and infant seals.

The diving response was found to be well developed in new-born lambs by Tchobroutsky, Merlet & Rey (1969) who

113

stimulated the afferent input via the laryngeal nerves. They suggested that stimulation of the glottis by the presence of amniotic fluid *in utero* might prevent respiratory movements of the fetus. However, fetal breathing is known to occur regularly in man and animals (Boddy & Dawes, 1975), and its movements are converted to gasping by asphyxia (Dawes, Fox & Richards, 1972). Remarkably, it has been found by Johnson, Robinson & Salisbury (1973) that laryngeal stimulation leading to apnoea and bradycardia responds to water or milk but not to saline! The issue of reflexes originating in the mouth and larynx of fetuses and their possible role in breathing movements remains in doubt.

Breathing in the new-born animal is enhanced by a variety of conditions. These include tactile stimuli and cooling, especially of the face (Barcroft, 1946; Dawes, 1965), umbilical ligation and hypoxia and hypercapnia (Chernick & Jansen, 1973). Stimulation of the aortic bodies of the fetal lamb by cyanide produces reflex femoral vasoconstriction and complex changes in heart rate, usually bradycardia, but stimulation of the carotid bodies has no effect even in concentrations higher than those required for aortic body excitation (Dawes *et al.*, 1968). Therefore, only the aortic bodies appear to be active in the fetus, and they probably represent the primary asphyxial signal.

7 MEDICAL IMPLICATIONS

> . . . when the heart beats more languidly, it is impossible to feel the pulse not only in the fingers but also in the wrist and in the temples, as in fainting, in hysterical manifestations, in asphyxia, in the more sickly and in those about to die (William Harvey, 1628).

The relevance of the asphyxial defence response to conditions such as near-drowning and birth asphyxia is clear. In others, such as sudden infant-death, it provides an attractive hypothesis. The diving response has been used recently as a clinical tool for the treatment of certain cardiac arrhythmias and as a simple test of autonomic function. There would appear to be considerable scope for such applications, particularly as they apply to asphyxia and cardiovascular medicine. These considerations may indicate further avenues for investigation into a variety of medical and surgical problems. We suggest that some of these instances in which apnoea, hypoventilation and asphyxia are prominent parts of the clinical syndrome may be better understood by recognition of potential similarities in underlying mechanisms to the phenomena of diving responses. The subject has been reviewed (Angell-James & Daly, 1969b; Daly & Angell-James, 1979; Daly, Angell-James & Elsner, 1979a, b; Gooden, 1982).

Near-drowning

Drowning is a common cause of death, accounting for about 8000 fatalities per year in the USA (Baker, 1954; Boucher, 1962; Modell, 1971). Death from drowning may result from asphyxia alone, in which case little or no water is found in the lungs at autopsy, or from the consequences of inhalation of water combined with asphyxia (Miles, 1962). Young people are particularly at risk. A review of drowning deaths in

115

Medical implications

Australia for the year 1970 showed that the greatest number of deaths occurred in the zero- to nine-year age group (Gooden, 1972). In this group the greatest proportion were less than four years of age. This age group is susceptible to death from many forms of accidents because the children place themselves unwittingly into situations of danger. Drowning is no exception to this pattern.

Schneider (1930) examined 318 American aviators who held their breaths in air for an average of 68 s. Rahn (1964) estimated on purely theoretical grounds that after a deep inspiration a man could hold his breath for only 2 min. Even professional breath-hold divers such as the Ama of Korea and Japan and the pearl divers of the Torres Strait archipelago usually do not remain submerged for longer than a minute. Maximum dive times are between 2 and 2.5 min (Scholander *et al.*, 1962a; Hong & Rahn, 1967). It might therefore be expected that an adult drowning victim would succumb under water within 1 or 2 min, when the desire to breathe would result in the inhalation of water.

However, in recent years reports have appeared which clearly demonstrate man's ability to survive immersion for periods considerably in excess of the usual breath-holding time (Table 7.1). Most of the victims were male, as are the majority of drowning cases. The longest durations were of young children submerged in ice-cold water with correspondingly low rectal temperatures. In the case described by Hunt (1974) recovery was rapid and involved a minimum of sophisticated medical intervention. In the case reported by Siebke *et al.* (1975) recovery was much slower and required considerably more complex treatment. A sodium bicarbonate infusion restored the pH to normal and a strong regular pulse was felt for the first time about an hour after admission and 1 h and 45 min after submersion. Pulmonary oedema was dispersed by ventilation with a volume-controlled respirator with a positive end-expiratory pressure of 10 cm H_2O. A review of cases of near-drowning in children has indicated a better than 90% chance of full neurological recovery in surviving children who were apnoeic and unconscious when pulled from the water (Pearn *et al.*, 1979). Nemiroff (1977) summarized the results of 12 surviving cases of immersion times of from 1 to 38 min. Apnoea, cyanosis, bradycardia and

116

Table 7.1. *Survival after prolonged submersion*

Age (years)	Duration of submersion (min)	Water temperature (°C)	Initial rectal temperature (°C)	References
5	40	0	24	Siebke *et al.* (1975)
5	30 (est.)	0–1	27	Hunt (1974)
5	22 (est.)	0	24	Kvittingen & Naess (1963)
2	20	5–7	30	Imbach, Kabus & Tönz (1975)
16 months	20 (est.)	11.4 (air temp.)	28	De Villota *et al.* (1973)
3	20 (est.)	–	27 (? rectal)	Ohlsson & Beckman (1964)
21	>17	Autumn, Melbourne, Australia	32	King & Webster (1964)
15	7–10 (est.)	Summer, Sydney, Australia	–	Warden (1967)
23	25	0	29	Sekar *et al.* (1980)
27	6	0	33	Sekar *et al.* (1980)

hypothermia were the initial clinical findings.

These and other cases demonstrate that under certain circumstances the human body can stand long underwater asphyxia. This fact raises two fundamental questions. Why did these victims not inhale large quantities of water into the lungs? How did the cerebral and cardiac tissue escape irreversible damage? The failure of water to penetrate to the lungs may result from the inhibition, either voluntarily or reflexly, of respiratory movements. King and Webster (1964) (Table 7.1) suggested that their patient who survived immersion for more than 17 min may have been unconscious on entering the water. No water was found in his trachea on admission to the hospital.

The apnoea that occurs upon sudden submersion in water appears to be reflex in origin. Some human subjects, when submerged but free to breathe through a snorkel, find it nearly impossible to take an inspiration (Hamilton & Mayo,

1944). Laryngospasm probably prevents penetration of water into the lungs after it has been inhaled through the nose or mouth (Colebatch & Halmagyi, 1962; Ohlsson & Beckman, 1964; Griffin, 1966). Reflex spasm of the glottis in response to water entering the larynx has been unequivocally demonstrated in experimental animals by Banting *et al.* (1938). This glottal spasm was shown to produce a water-tight seal and to be abolished by spraying the larynx with cocaine. Stimulation of the central end of the superior laryngeal nerve in man produces reflex arrest of breathing (Ogura & Lam, 1953) reminiscent of similar effects in experimental animals (Chapter 5). The true incidence of 'dry drowning' in which water fails to penetrate to the lungs, and death is said to result from asphyxia alone, remains obscure. Experimental drowning of dogs has shown that 20–35% of animals do not inhale water (Fainer, Martin & Ivy, 1951; Swann, 1956). Cot (1931) and Moritz (1944) estimated that 15% or less of human drowning victims do not aspirate water.

Immersion in cold water results in rapid heat loss from the surface of the body and a precipitous fall in core temperature (Michenfelder *et al.*, 1963; Keatinge, 1969; Mohri *et al.*, 1969). As the core temperature falls the metabolic rate of even the tissues that continue to be perfused, such as the brain, will be depressed. Cerebral metabolic rate in dogs was reduced by 55% when the brain temperature was lowered from 38 to 28 °C and the whole body oxygen uptake decreased by 49% for the same temperature change (Michenfelder & Theye, 1968). The small body mass and surface area to mass ratio of young children, and consequent more rapid cooling in cold water, is doubtless advantageous in near-drowning survival. Victims who are knocked unconscious immediately before or upon submersion may have a greater chance of survival due to loss of voluntary muscle movement and the absence of higher cortical influences that may tend to override the development of the diving response (Wolf *et al.*, 1965).

Cardiac activity may persist despite the appearance of cyanosis and complete absence of limb pulses (Kvittingen & Naess, 1963; Siebke *et al.*, 1975). Intense peripheral vasoconstriction with marked bradycardia and reduced cardiac output can produce this clinical picture. Under these circum-

118

stances the use of vasodilator drugs would be contra-indicated, since a precipitous fall in arterial blood pressure may result (Kvittingen & Naess, 1963).

Asystole may occur in the near-drowned victim. De Villota *et al.* (1973) remarked that the asystole they observed in their patient, who had fallen into cold water, could not be wholly explained by hypothermia. They estimated that their patient had a rectal temperature of 25°C when he was recovered. Rhythmic heart activity has been reported at temperatures as low as 20°C (Mohri *et al.*, 1969) and a slow sinus rhythm has been observed at 25°C (Edwards *et al.*, 1970). A pronounced diving response can trigger such an intense parasympathetic discharge that cardiac arrest results. Anaesthetized adults cooled to a temperature below 28°C in preparation for cardiac surgery show a sharp increase in the incidence of ventricular arrhythmias (Ross, 1957). Below 25°C ventricular fibrillation is common. Children, however, appear to be more resistant to this condition (Benazon, 1960) and this may play an important part in their survival of hypothermia associated with near-drowning.

Provided the cardiac tissue has not suffered irreversible anoxic damage, recovery of cardiac activity would be expected using external cardiac massage and artificial ventilation. In the presence of peripheral vasoconstriction the parenteral administration of cardiac stimulants may be ineffectual (Kvittingen & Naess, 1963; Siebke *et al.*, 1975). Pronounced bradycardia is probably a natural and protective response in the hypothermic near-drowned victim and, if the rate is regular and sustained, it is not an indication *per se* for the administration of cardiac stimulants or atropine.

External rewarming is a valuable part of the treatment (Hunt, 1974). More recently the use of warm fluid by intravenous infusion, gastrointestinal irrigation and dialysis has been advocated (Sekar *et al.*, 1980). It is essential, however, to ensure that the oxygen supply to the body tissues keeps pace with their increasing metabolic demands as their temperature rises. The effective transport of oxygen from the lungs to the circulation is critical. Because of the efficacy of lung inflation in reflexly abolishing vagal bradycardia, it may play an important additional role when mouth-to-mouth resuscitation is employed in efforts to revive patients in cardiac arrest.

Pulmonary oedema is a not uncommon problem associated with near-drowning both in sea-water and in fresh water (Modell, 1971). Fortunately, the use of volume-controlled respirators with positive end-expiratory pressure are most effective in reversing this complication (Modell *et al.*, 1974; Siebke *et al.*, 1975). Arterial blood gas and pH estimations, corrected for temperature (White, 1981), would be valuable for gauging the pace of rewarming and artificial ventilation. It is possible that the renal ischaemia evoked by the diving response may be sufficient to result in hypoxic damage of the kidneys. The two cases reported by Grausz, Amend & Earley (1971) may be examples of such a mechanism.

Sudden underwater deaths

Sudden loss of consciousness underwater in otherwise normal healthy people is well documented (Craig, 1961*a*; Dumitru & Hamilton, 1963). Two basic theories, respiratory and neurogenic, have been proposed to explain these deaths. Craig (1961*a*, *b*) proposed that swimmers who hyperventilate before diving lower the partial pressure of carbon dioxide in the arterial blood but produce little effect upon the partial pressure of oxygen. During underwater swimming the oxygen tension falls below a level necessary for consciousness before the carbon dioxide reaches a level sufficient to stimulate the diver to surface and breathe. Dumitru & Hamilton (1963) proposed a similar mechanism, but also suggested that the hypocapnia produced by hyperventilation could result in cerebral vasoconstriction and hence cerebral hypoxia.

Breath-hold divers ascending from considerable depth should be particularly vulnerable to this latter hypoxic mechanism since the alveolar oxygen tension falls precipitously during ascent (Schaefer & Carey, 1962; Paulev, 1968*b*). US Navy submarine escape instructors were studied after ascent from a depth of 27 m. Alveolar P_{O_2} fell during ascent to reach values as low as 30 mmHg (Schaefer & Carey, 1962). One diver became unconscious for a short time on reaching the surface.

Sudden loss of consciousness and death may occur without preceding hyperventilation (Gardner, 1942). Glaister (1947) suggested that the impingement of cold water on the nasal and post-nasal mucous membranes after sudden immersion might trigger a cardio-inhibitory reflex. Gjone (1961) believed that some victims of sudden immersion died as a result of reflex cardiac arrest immediately upon submersion. Wolf (1964) extended this theory by proposing that overactivity of the diving response could lead to serious arrhythmias, cardiac arrest and death. He and his colleagues demonstrated the influence of the emotional state of the subject on the cardiac response to face immersion (Wolf *et al.*, 1965). Fear appeared to augment the diving bradycardia. Craig (1963) remarked that a very poor swimmer who usually avoided immersion further than the neck developed a pronounced bradycardia during diving. Breath-holding both in air and underwater has been shown to be associated with a variety of arrhythmias (Lamb, Dermskian & Sarnoff, 1958; Olsen *et al.*, 1962*a*).

Such a neurogenic mechanism may have been the cause of death in the famous 'Brides in the Bath' murders (Watson, 1922). Smith was believed to have caused the death of each of his three victims by submerging her head while she was taking a bath. In order to test this possibility one of the detectives on the case persuaded a young lady of his acquaintance, who was a practised swimmer and accustomed to having her head under water, to sit in a bath which was filled to the height found in one of the murders. A. F. Neil described the test thus:

> It was decided to test sudden immersion, so from the ankle, I lifted up her legs very suddenly. She slipped under easily, but to me, who was closely watching, she seemed to make no movement. Suddenly, I gripped her arm, it was limp. With a shout I tugged at her armpit and raised her head above the water. It fell over to one side. She was unconscious. For nearly half an hour my detectives and I worked away at her with artificial respiration and restoratives. The lady explained afterwards that immediately she fell back with her legs held in the air, the water rushed into her mouth and up her nostrils, making her unconscious (Crew, 1933).

Medical implications

Angell-James & Daly (1975) suggested that stimulation with water of the regions innervated by trigeminal or superior laryngeal nerves results in inhibition of carotid body chemoreceptor drive. Hence the victim loses consciousness before being driven to the surface by the compelling urge to breathe.

Asphyxia in new-born humans

Asphyxia, usually of brief duration, occurs to some degree as a result of the delivery process in all births (James *et al.*, 1958). Vigorous infants demonstrate a wide range of umbilical artery oxygen saturation after delivery, some crying lustily with low oxygen in their arterial blood. The protective reactions against asphyxia shown by fetal and new-born mammals resemble in some respects the diving response of marine mammals and therefore may confer a potential physiological advantage by conserving oxygen and maintaining preferentially the cerebral blood flow (Chapter 5). Any treatment which antagonizes the cardiovascular components of this response, either by vagal blockade or by blocking the reflex vasoconstriction while the apnoea persists, is likely to have a serious deleterious effect on the patient by eliminating a natural defence mechanism. Thus, treatment of the severe asphyxia must include reversal of the hypoxic and apnoeic state by restoration of adequate gas exchange (Cross, 1966).

New-born mammals are especially sensitive to the application of certain liquids to the laryngeal mucosa and to electrical stimulation of the afferent pathways for the reflex, the superior laryngeal nerves, which results in apnoea, bradycardia, hypertension and redistribution of a reduced cardiac output to the brain and myocardium (Harding *et al.*, 1976). The apnoea and bradycardia may be prolonged, on rare occasions even leading to death (Downing and Lee, 1975; Johnson, Salisbury & Storey, 1975). Pathological ischaemia may occur in some organs (Chapter 6).

Sudden infant death syndrome

The clinical definition of sudden infant death syndrome (SIDS) is 'the death of a child aged between two weeks and two years, who was thought to be in good health, or whose terminal illness appeared to be so mild that the possibility of a fatal outcome was not expected' (Beal, 1972). However, about 15% of these infants are found at post-mortem examination to have pathological conditions that can explain death.

The incidence of this syndrome has been reported as being 1.2–3 per 1000 live births (Valdes-Dapena, 1970). The majority of victims die suddenly, silently and while asleep (Wedgwood, 1972). There is a higher incidence in premature babies and during the winter months (Valdez-Dapena, 1967). The etiology of this syndrome remains a mystery. Suggested causes are manifold, and the reader is referred for reviews to works edited by Bergman, Beckwith & Ray (1970), Camps & Carpenter (1972) and La Veck (1972). Some aspects of the SIDS bring to mind the responses of diving animals and the control mechanisms which govern them. The relative insensitivity of respiratory regulation and apnoea combined with chemoreceptor and nasopharyngeal stimulation are, in some instances, common to both.

Considerable interest has been directed recently towards the possibility that some form of autonomic dysfunction might be involved. Wedgwood (1972) pointed out that this age-group is characterized by marked instability of the cardiorespiratory reflexes. He postulated that an acute period of respiratory obstruction might be sufficient to trigger cardiac arrest with ensuing death in susceptible children. Naeye *et al.* (1976) found carotid body abnormalities in many SIDS victims. Upper respiratory tract infection with nasopharyngeal oedema or narrowing of the nasal passages might be predisposing factors (Shaw, 1968; Cross & Lewis, 1971). Some infants appear to be almost completely unable to breathe through the mouth and this condition may persist until six months of age. Shaw (1970) has estimated that 30% of infants between birth and six months of age will not breathe through their mouths. Adelson (1953) proposed that instantaneous deaths might be due to an inhibitory reflex mechanism leading to arrest of the heart. The triggering factor may be so minor as to leave no visible evidence at autopsy. The immature hearts of goats, pigs and puppies have been shown to be more susceptible to arrhythmias than the adult organ (Preston, McFadden & Moe, 1959).

French, Morgan & Guntheroth (1972) examined the cardiac and respiratory responses of neonatal monkeys to nasal occlusion or face immersion in cold water (14°C). Episodes of prolonged breath-holding occurred in younger monkeys and they persisted beyond the stimulus period. In one episode the animal recovered spontaneously while in others resuscitation was required. These findings are consistent with the observation that some victims of sudden infant death have had periods of apnoea before death (Steinschneider, 1972; Guilleminault *et al.*, 1975).

Further evidence of the sensitivity of cardiorespiratory responses in young children was obtained by Steinschneider (1970) who studied the effect of bottle feeding. Among the children studied was a 41-day-old premature infant who had been referred because of recurrent apnoeic periods. Breath-holding and marked bradycardia developed while a bottle was sucked. The heart rate decreased from 160 beats/min to 80 beats/min. In about 1 min the infant became pale and still remained apnoeic. When the bottle was removed, breathing recommenced and the heart rate recovered rapidly.

One possible mechanism for these phenomena relates to sensory inputs affecting respiration and the cardiovascular system, particularly in new-born animals, from receptors in the upper airways innervated by the trigeminal and superior laryngeal nerves. The larynx is well endowed with receptors responding to mechanical and chemical stimuli (Boushey *et al.*, 1974) and their stimulation causes reflex apnoea, bradycardia and hypertension. Quantitatively the responses depend very much on maturity. In the new-born mammal, prolonged apnoea, sometimes leading to death, can be produced. That response is only transient in the adult animal (Johnson, Dawes & Robinson, 1972; Johnson *et al.*, 1973). In infants gastro-oesophageal reflux sometimes causes respiratory arrest and cyanosis (Leape *et al.*, 1977). Receptors in the nose and on the face of new-born animals also cause reflex apnoea and bradycardia when stimulated (Tchobroutsky *et al.*, 1969), sometimes leading to death when the stimulus is sustained (Wealthal, 1975). Similarly, nasopharyngeal suc-

124

tion in new-born infants can produce apnoea and cardiac arrhythmias (Cordero & Hon, 1971).

Another avenue of investigation has focused on the control of respiration. The sensitivity of the respiratory response to increasing the inspired carbon dioxide concentration is as great in full-term infants as in adults (Cross, Hooper & Oppe, 1953; Avery *et al.*, 1963). However, in premature infants the sensitivity of the response is considerably reduced (Rigatto, Brady & Verduzco, 1975; Frantz *et al.*, 1976). A deficient response of alveolar ventilation to carbon dioxide is also seen in premature infants with episodic apnoea (Rigatto & Brady, 1972) and in full-term infants with central hypoventilation during quiet sleep (Shannon *et al.*, 1976; Shannon, Kelley & O'Connel, 1977). Thus immaturity of the carotid bodies is unlikely to be the explanation for the apparent lack of chemoreceptor reflex activity in infants with periodic breathing and apnoeic spells (Rigatto & Brady, 1972). A central depressant effect of hypoxia and of reduced sensitivity to carbon dioxide, related to prematurity (Rigatto & Brady, 1972; Rigatto *et al.*, 1975), is more likely.

In summary, these findings suggest that respiration can be depressed in the new-born infant by central mechanisms, especially hypoxia, and reflexly through stimulation of upper-airway receptors, this leading to asphyxia despite the presence of a peripheral chemoreceptor drive. During an apnoeic episode an initial bradycardia would be expected both because of loss of the normal rhythmic central inspiratory drive and because of decreased pulmonary stretch receptor activity (Anrep *et al.*, 1936*a*, *b*; Daly & Scott, 1958; Daly & Angell-James, 1975; Haymet & McCloskey, 1975). These conditions of apnoeic asphyxia are favourable to the production of the chemoreceptor-induced bradycardia (Angell-James & Daly, 1973; Daly *et al.*, 1978). This situation could be enhanced or initiated by regurgitation of stomach contents (Leape *et al.*, 1977), particularly when it contains cow's milk (Parish *et al.*, 1960; Henschel & Coates, 1974). If such an event occurred during sleep it could be expected to produce apnoea, cyanosis and even airway obstruction initiated by laryngospasm (Szereda-Przestaszewska & Widdicombe, 1973; Leape *et al.*, 1977).

Potentially hazardous conditions

Inhalation of noxious or foreign gases and liquids may be especially hazardous in the presence of chemoreceptor stimulation, resulting in unexpectedly severe cardiorespiratory reactions. Among these influences are smoke and aerosols (Bass, 1970; Daly & Taton, 1979). Chronic alveolar hypoventilation resulting in arterial hypoxaemia, sometimes associated with respiratory distress, sleep apnoea syndrome (Guilleminault *et al.*, 1977), airway aspiration in quadraplegic patients (Frankel, Mathias & Spalding, 1975) and other similar conditions present potential hazards.

A superficial similarity between the vascular adjustments that occur in response to haemorrhagic shock and those of diving asphyxia has been noted (Scholander, 1963; Elsner *et al.*, 1966*a*). Although a general redistribution of circulation tending to favour coronary and cerebral perfusion can be discerned (Gregg, 1962; Chien, 1967), a cautious interpretation is required, because of the variety of shock conditions and reactions to them (Rushmer, Van Citters & Franklin, 1962).

The diving response in clinical applications

The present and possible future applications of the diving response in clinical medicine have recently been reviewed (Gooden, 1982). Face immersion in water at 2 °C was found to be useful in treatment of paroxysmal atrial tachycardia in some patients (Wildenthal *et al.*, 1975). Four of their patients had a history of attacks, while three others had no previous history of heart disease. Face immersion breath-holding converted the paroxysmal atrial tachycardia to sinus rhythm within 15 to 35 s in the seven patients. Two patients failed to respond at the first testing but a repeat test a few minutes later converted the rhythm. The face immersion procedure is not associated with gross changes in arterial blood pressure and may offer a useful adjunct to carotid sinus massage and intravenous infusion of vasopressors for the treatment of this arrhythmia. Conventional therapy carries definite risks (Scherf & Bornemann, 1966; Benaim, 1972; Cohen, 1972), and is not always successful.

Caution should be observed in the use of the face immersion procedure, since some patients may be particularly sensitive to this stimulus. Whayne & Killip (1967) reported the case of a 57-year-old man who developed heart block following aortic valve replacement. While washing his face with a cold, wet washcloth he suddenly fell unconscious. Pickering & Bolton-Maggs (1975) successfully used water at 15 °C in the treatment of supraventricular tachycardia associated with myocardial infarction. Because of earlier work which showed that breath-holding or diving triggered arrhythmias in some subjects, Wildenthal *et al.* (1975) considered that the face immersion procedure should not be used in these patients. However Gooden *et al.* (1978) subjected 20 patients recovering from myocardial infarction to face immersion with no evidence of pathological arrhythmias.

The diving response is unique in that it simultaneously elicits increased activity in specific components of both divisions of the autonomic nervous system, the vagus nerve and some specific vasoconstrictor nerves. Unlike other tests of autonomic function such as prolonged tilting (Johnson & Spalding, 1974), vasopressor tests (Smyth, Sleight & Pickering, 1969) and the Valsalva manoeuvre (Sharpey-Schafer, 1965), arterial blood pressure during the diving response remains normal or is only moderately elevated (Heistad *et al.*, 1968; Campbell *et al.*, 1969*a*). Bennett, Hosking & Hampton (1976) examined the cardiovascular responses of diabetic patients to a face immersion test.

The pronounced effect of lung inflation on excitation of the pulmonary stretch receptors, which results in increased heart rate, is discussed in Chapter 5. Stimulation of that reflex mechanism is likely to contribute to the effectiveness of mouth-to-mouth breathing as a technique for cardiopulmonary resuscitation. Thus the act of lung inflation may have a possible second function in reversing cardiac arrest or elevating the heart rate in addition to its role in supporting gas exchange (Daly *et al.*, 1979*b*).

Myocardial ischaemia

The treatment of ischaemia and infarction of the human heart requires that the blood and oxygen supply to the heart should

127

be facilitated or the work load on the heart should be reduced, or both. It has been shown that the diving response evoked in dogs during trained snout immersion produces bradycardia and a considerable reduction in myocardial oxygen consumption (Gooden *et al.*, 1974). Kjekshus & Blix (1977) suggested that myocardial infarction could be treated by induced bradycardia with reference to the decreased cardiac work of diving seals.

Several approaches along these lines have recently been explored in experimental animals and patients with acute coronary occlusion (reviewed by Guyton & Daggett, 1976; Hillis & Braunwald, 1977). Braunwald *et al.* (1967) used electrical stimulation of the carotid sinus to relieve angina pectoris in coronary patients. Similarly, beta-receptor blockade alleviates angina pectoris and probably exerts its effect on the ischaemic myocardium mainly through its ability to reduce heart rate (Mueller *et al.*, 1974). However, the effect of beta-blockade on infarct size in human patients is unclear (Frishman, 1980). Other work in experimental animals has indicated that the size of a myocardial infarction can be minimized by maintenance of a relatively slow heart rate (Redwood, Smith & Epstein, 1972; Shell & Sobel, 1973; Myers *et al.*, 1974; Kjekshus *et al.*, 1981*a*, *b*).

Autonomic function is often disturbed following infarction (Pantridge, Webb & Adgey, 1975) and some work has been directed at an examination of the parasympathetic drive to the heart in these patients. Ryan *et al.* (1976) reported that myocardial infarction patients, 3–18 months after their cardiac incident, showed a smaller bradycardial response to face immersion than normal control subjects. Gooden *et al.* (1978), using water at 20°C for face immersion, reported that patients who had suffered a myocardial infarction within 3–13 days previously had greater bradycardia than age-matched control patients without ischaemic heart disease. Hence the infarct patients demonstrated a proclivity for reflex bradycardia (Fig. 7.1). Wolf *et al.* (1965) postulated that nerve endings in coronary vessels and the myocardium might be stimulated by ischaemia, giving rise to cardiovascular adjustments designed to conserve oxygen in the face of an impaired pumping action of the heart.

128

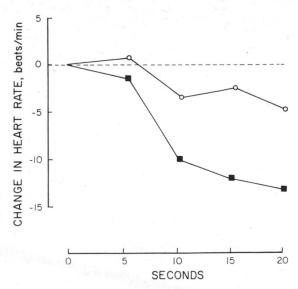

Fig. 7.1. Heart rate change in human subjects during face immersion experiments. Circles, controls; squares, myocardial infarction patients. (Redrawn from Gooden *et al.*, 1978.)

REFERENCES

Ackerman, R. A. & White, F. N. (1979). Cyclic carbon dioxide exchange in the turtle *Pseudemys scripta*. *Physiol. Zool.* **52**, 378–89.

Adamsons, K., Behrman, R., Dawes, G. S., Dawkins, M. J. R., James, L. S. & Ross, B. B. (1963). The treatment of acidosis with alkali and glucose during asphyxia in foetal rhesus monkeys. *J. Physiol.* **169**, 679–89.

Adelson, L. (1953) Possible neurological mechanisms responsible for sudden death with minimal anatomical findings. *J. Forensic Med.* **1**, 39–45.

Allen, F. M. (1928). Effect on respiration, blood pressure, and carotid pulse of various inhaled and insufflated vapors when stimulating one cranial nerve and various combinations of cranial nerves. I. Branches of the trigeminus affected by these stimulants. *Am. J. Physiol.* **87**, 319–25.

Allen, F. M. (1937). Local asphyxia and temperature changes in relation to gangrene and other surgical problems. *Trans. Assn. Am. Phys.* **52**, 189–94.

Allen, F. M. (1938*a*). Resistance of peripheral tissues to asphyxia at various temperatures. *Surg. Gynec. Obst.* **67**, 746–51.

Allen, F. M. (1938*b*). The tourniquet and local asphyxia. *Am. J. Surg.* **41**, 192–200.

Allison, D. J. & Powiss, D. A. (1971). Adrenal catecholamine secretion during stimulation of the nasal mucous membrane in the rabbit. *J. Physiol.* **217**, 327–40.

Anand, B. K., Chhina, G. S. & Singh, B. (1961). Studies on Shri Ramanand Yogi during his stay in an air-tight box. *Ind. Jour. Med. Res.* **49**, 82–9.

Andersen, H. T. (1959). Depression of metabolism in the duck during diving. *Acta Physiol. Scand.* **46**, 234–9.

Andersen, H. T. (1961). Physiological adjustments to prolonged diving in the American alligator *Alligator mississippiensis*. *Acta Physiol. Scand.* **53**, 23–45.

Andersen, H. T. (1963*a*). Factors determining the circulatory adjustments to diving. I. Water immersion. *Acta Physiol. Scand.* **58**, 173–85.

Andersen, H. T. (1963*b*). Factors determining the circulatory adjustments to diving. II. Asphyxia. *Acta Physiol. Scand.* **58**, 186–200.

Andersen, H. T. (1963*c*). The reflex nature of the physiological adjustments to diving and their afferent pathway. *Acta Physiol. Scand.* **58**, 263–73.

Andersen, H. T. (1966). Physiological adaptations in diving vertebrates. *Physiol, Rev.* **46**, 212–43.

References

Andersen, H. T. (Ed.) (1969). *The Biology of Marine Mammals*, 511 pp. New York: Academic Press.

Andersen, H. T. & Blix, A. S. (1974). Pharmacological exposure of components in the autonomic control of the diving reflex. *Acta Physiol. Scand.* **90**, 381–6.

Andersen, H. T., Hustvedt, B. E. & Lövö, A. (1965). Acid–base changes in diving ducks. *Acta Physiol. Scand.* **63**, 128–32.

Angell-James, J. E. & Daly, M. de B. (1969a). Cardiovascular responses in apnoeic asphyxia: role of arterial chemoreceptors and the modification of their effects by a pulmonary vagal inflation reflex. *J. Physiol.* **201**, 87–104.

Angell-James, J. E. & Daly, M. de B. (1969b). Nasal reflexes. *Proc. Roy. Soc. Med.* **62**, 1287–93.

Angell-James, J. E. & Daly, M. de B. (1972a). Reflex respiratory and cardiovascular effects of stimulation of receptors in the nose of the dog. *J. Physiol.* **220**, 673–96.

Angell-James, J. E. & Daly, M. de B. (1972b). Some mechanisms involved in the cardiovascular adaptations to diving. *Soc. Exp. Biol. Symp.* **26**, 313–41.

Angell-James, J. E. & Daly, M. de B. (1973). The interaction of reflexes elicited by stimulation of carotid body chemoreceptors and receptors in the nasal mucosa affecting respiration and pulse interval in the dog. *J. Physiol.* **229**, 133–49.

Angell-James, J. E. & Daly, M. de B. (1975). Some aspects of upper respiratory tract reflexes. *Acta Oto-Laryngol.* **79**, 242–51.

Angell-James, J. E. & Daly, M. de B. (1978). The effects of artificial lung inflation on reflexly induced bradycardia associated with apnoea in the dog. *J. Physiol.* **274**, 349–66.

Angell-James, J. E., Daly, M. de B. & Elsner, R. (1978). Arterial baroreceptor reflexes in the seal and their modification during experimental dives. *Am. J. Physiol.* **234**, H730–9.

Angell-James, J. E., Elsner, R. & Daly, M. de B. (1981). Lung inflation: Effects on heart rate, respiration and vagal activity in seals. *Am. J. Physiol.* **240**, H190–8.

Anrep, G. V., Pascual, W. & Rössler, R. (1936a). Respiratory variations of the heart rate. I. The reflex mechanism of the respiratory arrhythmia. *Proc. Roy. Sc. Ser. B* **119**, 191–217.

Anrep, G. V., Pascual, W. & Rössler, R. (1936b). Respiratory variations of the heart rate. II. The central mechanism of the respiratory arrhythmia and the interrelations between the central and reflex mechanisms. *Proc. Roy. Soc. Ser. B* **119**, 218–30.

Ashwell-Erickson, S. & Elsner, R. (1981). The energy cost of free existence for Bering Sea harbor and spotted seals. In: *The Eastern Bering Sea Shelf: Oceanography and Resources*, ed. D. W. Hood and J. A. Calder, pp. 879–99. Washington, DC: US Dept. of Commerce 2.

Asmussen, E. & Kristiansson, N. G. (1968). The 'diving bradycardia' in exercising man. *Acta Physiol. Scand.* **73**, 527–35.

Augee, M. L., Elsner, R., Gooden, B. A. & Wilson, P. R. (1971).

References

Respiratory and cardiac responses of a burrowing animal, the echidna. *Respir. Physiol.* **11**, 327–34.

Avery, M. E., Chernick, V., Dulton, R. E. & Permutt, S. (1963). Ventilatory response to inspired carbon dioxide in infants and adults. *J. Appl. Physiol.* **18**, 895–903.

Bainton, C. R., Elsner, R. & Matthews, R. C. (1973). Inhaled CO_2 and progressive hypoxia: ventilatory response in a yearling and a newborn harbor seal. *Life Sci.* **12**, Part II, 527–33.

Baker, A. Z. (1954). Drowning and swimming. *Practitioner* **172**, 655–9.

Baker, H. de C. (1956). Ischaemic necrosis in the rat liver. *J. Pathol. Bacteriol.* **71**, 135.

Bamford, O. S. & Jones, D. R. (1974). On the initiation of apnoea and some cardiovascular responses to submergence in ducks. *Respir. Pysiol.* **22**, 199–216.

Banting, F. G., Hall, G. E., Janes, J. M., Leibel, B. & Lougheed, D. W. (1938). Physiological studies in experimental drowning. *Canad. Med. Assn. J.* **39**, 226–8.

Barcroft, H. & Swan, H. J. C. (1953). *Sympathetic Control of Human Blood Vessels.* London: Edward Arnold & Co.

Barcroft, J. (1920). Physiological effects of insufficient oxygen supply. *Nature* **106**, 125–9.

Barcroft, J. (1946). *Researches on Prenatal Life.* Oxford: Blackwell.

Bartholomew, G. A. (1954). Body temperature and respiratory and heart rates in the northern elephant seal. *J. Mammal.* **35**, 211–19.

Bartholomew, G. A., Bennett, A. F. & Dawson, W. R. (1976). Swimming, diving, and lactate production of the marine iguana, *Amblyrhynchus cristatus. Copeia* **4**, 709–20.

Bass, M. (1970). Sudden sniffing death, *J. Am. Med. Assn.* **212**, 2075–9.

Bauer, D. J. (1937). The slowing of the heart rate produced by clamping the umbilical cord in the foetal sheep. *J. Physiol.* **90**, 25P–27P.

Bauer, D. J. (1938). The effect of asphyxia upon the heart rate of rabbits at different ages. *J. Physiol* **93**, 90–103.

Beal, S. (1972). Sudden infant death syndrome. *Med. J. Aust.* **2**, 1223–9.

Behrisch, H. W. & Elsner, R. (1980a). Molecular adaptations to the diving habit. Fructose diphosphatase from a diver, the harbor seal. *Comp. Biochem. Physiol.* **66B**, 123–7.

Behrisch, H. W. & Elsner, R. (1980b). Molecular adaptations to the diving habit. On the role of glycerol in the ischemic seal kidney. *Comp. Biochem. Physiol.* **66B**, 403–7.

Behrman, R. E., Lees, M. H., Peterson, E. N., de Lannoy, C. W. & Seeds. A. E. (1970). Distribution of the circulation in the normal and asphyxiated fetal primate. *Am. J. Obstet. Gynecol.* **108**, 956–69.

Belkin, D. A. (1962). Anaerobiosis in diving turtles. *Physiologist* **5**, 105–10.

Belkin, D. A. (1963). Anoxia: Tolerance in reptiles. *Science* **139**, 492–3.

Benaim, M. E. (1972). Asystole after verapamil. *Brit. Med. J.* **2**, 169–70.

Benazon, D. (1960). The experimental and clinical use of profound hypothermia. *Anaesthesia* **15**, 134–45.

Bennett, T., Hosking, D. J. & Hampton, J. R. (1976). Cardiovascular reflex responses to apnoeic face immersion and mental stress in diabetic subjects. *Cardiovascular Res.* **10**, 192–9.

Bergman, A. B., Beckwith, J. B. & Ray, C. G. (1970). *Sudden Infant Death Syndrome*, 248 pp. Proc. of Second Int. Conf. on Causes of Sudden Death in Infants. Seattle: Univ. of Washington Press.

Bergman, S. A., Campbell, J. K. & Wildenthal, K. (1972). 'Driving reflex' in man: its relation to isometric and dynamic exercise. *J. Appl. Physiol.* **33**, 27–31.

Bernard, C. (1865). *Introduction à l'Etude de la Médecine Expérimentale*, trans, by H. C. Greene, 1927. Reprinted by Dover Publications, Inc., New York, 1957.

Bert, P. (1870). *Leçons Sur la Physiologie Comparée de la Respiration*, pp. 526–53. Paris: Baillière.

Birkeland, S., Vogt, A., Krog, J. & Semb, C. (1959). Renal circulatory occlusion and local cooling. *J. Appl. Physiol.* **14**, 227–32.

Biscoe, T. J., Purves, M. J. & Sampson, S. R. (1970). The frequency of nerve impulses in single carotid body chemoreceptor afferent fibres recorded *in vivo* with intact circulation. *J. Physiol.* **208**, 121–31.

Black, A. M. & Torrance, R. W. (1971). Respiratory oscillations in chemoreceptor discharge in the control of breathing. *Respir. Physiol.* **13**, 221–37.

Blair, D. A., Glover, W. E., Greenfield, A. D. M. & Roddie, I. C. (1959). Excitation of cholinergic vasodilator nerves to human skeletal muscles during emotional stress. *J. Physiol.* **148**, 633.

Blair, D. A., Glover, W. E. & Roddie, I. C. (1961). Vasomotor responses in the human arm during leg exercise. *Circulation Res.* **9**, 264–74.

Blessing, M. H. (1972). Studies on the concentration of myoglobin in the sea-cow and porpoise. *Comp. Biochem. Physiol.* **41A**, 475–80.

Blessing, M. H. & Hartschen, A. (1969). Beitrag zur Anatomie des Cavasphincters des Seehundes (*Phoca vitulina* L.). *Anat. Anz.* **124**, 105–12.

Blessing, M. H. & Hartschen-Niemeyer, E. (1969). Uber den Myoglobingehalt der Herz und Skelettmuskulatur insbesondere einiger mariner Säuger. *Z. Biol.* **116**, 302–13.

Blix, A. S. (1975). The importance of asphyxia for the development of diving bradycardia in ducks. *Acta Physiol. Scand.* **95**, 41–5.

Blix, A. S. (1976). Metabolic consequences of submersion asphyxia in mammals and birds. *Biochem. Soc. Symp.* **41**, 169–78.

Blix, A. S. & Folkow, B. (1983). Cardiovascular responses to diving in mammals and birds. In: *Handbook of Physiology, The Cardiovascular System*, vol. 4. Washington, DC: Am. Physiol. Soc., in press.

Blix, A. S. & Steen, J. B. (1979). Temperature regulation in newborn polar homeotherms. *Physiol. Rev.* **59**, 285–304.

Blix, A. S., Gautvik, E. L. & Refsum, R. (1974). Aspects of the relative roles of peripheral vasoconstriction and vagal bradycardia in the establishment of the diving reflex in ducks. *Acta Physiol. Scand.* **90**, 381–6.

Blix, A. S., Lundgren, O. & Folkow, B. (1975). The initial cardiovascular responses in the diving duck. *Acta Physiol. Scand.* **94**, 539–41.

Blix, A. S., Kjekshus, J. K., Enge, I. & Bergan, A. (1976*a*). Myocardial blood flow in the diving seal. *Acta Physiol. Scand.* **96**, 277–80.

Blix, A. S., Wennergren, G. & Folkow, B. (1976*b*). Cardiac receptors in ducks – a link between vasoconstriction and bradycardia during diving. *Acta Physiol. Scand.* **97**, 13–19.

Blix, A. S., Elsner, R. & Kjekshus, J. K. (1983). Cardiac output and its distribution through capillaries and A–V shunts in diving seals. *Acta Physiol. Scand.* in press.

Boddy, D. & Dawes, G. S. (1975). Fetal breathing. *Brit. Med. Bull.* **31**, 3–7.

Boucher, C. A. (1962). Drowning. *Monthly Bull. Minist, Health* (London) **21**, 114–17.

Boushey, H. A., Richardson, P. S., Widdicombe, J. G. & Wise, J. C. M. (1974). The response of laryngeal afferent fibres to mechanical and chemical stimuli. *J. Physiol.* **240**, 153–76.

Bove, A. A., Lynch, P. R., Connell, J. V. & Harding, J. M. (1968). Diving reflex after physical training. *J. Appl. Physiol.* **25**, 70–2.

Boyle, R. (1670). New pneumatical experiments about respiration. *Phil. Trans. Roy. Soc.* **5**, 2011–31.

Bradley, S. E. & Bing, R. J. (1942). Renal function in the harbor seal (*Phoca vitulina* L.) during asphyxial ischemia and pyrogenic hyperemia. *J. Cell. Comp. Physiol.* **19**, 229–37.

Bradley, S. E., Mudge, G. H. & Blake, W. D. (1954). The renal excretion of sodium, potassium and water by the harbor seal (*Phoca vitulina* L.): effect of apnea; sodium, potassium and water loading: pitressin and mercurial diuresis. *J. Cell. Comp. Physiol.* **43**, 1–22.

Braunwald, E. (1971). Control of myocardial oxygen consumption: Physiological and clinical considerations. *Am. J. Cardiol.* **27**, 416–32.

Braunwald, E., Epstein, S. E., Glick, G., Wechsler, A. & Braunwald, N. S. (1967). Relief of angina pectoris by electrical stimulation of the carotid sinus. *N. Engl. J. Med.* **277**, 1278–83.

Brick, I. (1966). Circulatory responses to immersing the face in water. *J. Appl. Physiol.* **21**, 33–6.

Brinkman, C. R., Mofid, M. & Assali, N. S. (1974). Circulatory shock in pregnant sheep. *Am. J. Obstet. Gynecol.* **118**, 77–90.

Brodie, T. G. & Russell, A. E. (1900). On reflex cardiac inhibition. *J. Physiol.* **26**, 92–106.

Bron, K. M., Murdaugh, H. V., Jr, Millen, J. E., Lenthall, R., Raskin, P. & Robin, E. D. (1966). Arterial constrictor response in a diving mammal. *Science* **153**, 540–3.

Bruner, J. M. (1970). Time, pressure, and temperature factors in the safe use of the tourniquet. *Hand* **2**, 39–42.

Bryan, R. M. & Jones, D. R. (1980*a*). Cerebral energy metabolism in diving and non-diving birds during hypoxia and apnoeic asphyxia. *J. Physiol.* **299**, 323–36.

Bryan, R. M. & Jones, D. R. (1980*b*). Cerebral energy metabolism in mallard ducks during apneic asphyxia: the role of oxygen conservation. *Am. J. Physiol.* **239**, R352–7.

134

Bullard, R. W., David, G. & Nichols, C. T. (1960). The mechanisms of hypoxic tolerance in hibernating and non-hibernating mammals. *Bull. Mus. Comp. Zool.* **124**, 321–36.

Burch, G. E. & Giles, T. D. (1970). A digital rheoplethysmographic study of the vasomotor response to 'simulated diving' in man. *Cardiology* **55**, 257–71.

Burow (1838). Ueber das Gefassystem der Robben. *Arch. Anat., Physiol. wissenschaftliche* Med. Heft ii, 230–58.

Butler, P. J. & Jones, D. R. (1968). Onset of and recovery from diving bradycardia in ducks. *J. Physiol.* **196**, 255–72.

Butler, P. J. & Jones, D. R. (1982). The comparative physiology of diving in vertebrates. *Advances in Comp. Physiol. Biochem.*, **8**, 179–364.

Butler, P. J. & Taylor, E. W. (1973). The effect of hypercapnic hypoxia accompanied by different levels of lung ventilation on heart rate in the duck. *Respir. Physiol.* **19**, 176–87.

Butler, P. J. & Woakes, A. J. (1979). Changes in heart rate and respiratory frequency during natural behavior of ducks with particular reference to diving. *J. Exper. Biol.* **79**, 283–300.

Bynum, T. E., Ruoff, P. A. & Rickert, J. (1970). The smoke reflex in rabbits. An example of autonomic control of cardiorespiratory function. *Cond. Reflex.* **5**, 233–40.

Calder, W. A. (1969). Temperature relations and underwater endurance of the smallest homeothermic diver, the water shrew. *Comp. Biochem. Physiol.* **30**, 1075–82.

Caldwell, P. R. B. & Wittenberg, B. A. (1974). The oxygen dependency of mammalian tissues. *Am. J. Med.* **57**, 447–52.

Campbell, A. G. M., Dawes, G. S., Fishman, A. P. & Hyman, A. I. (1967). Regional redistribution of blood flow in the mature fetal lamb. *Circulation Res.* **21**, 229–35.

Campbell, L. B., Gooden, B. A. & Horowitz, J. D. (1969a). Cardiovascular responses to partial and total immersion in man. *J. Physiol.* **202**, 239–50.

Campbell, L. B., Gooden, B. A., Lehman, R. G. & Pym, J. (1969b). Simultaneous calf and forearm blood flow during immersion in man. *Aust. J. Exp. Biol. Med. Sci.* **47**, 747–54.

Campbell, K. B., Rhode, E. A., Cox, R. H., Hunter, W. C. & Noordergraaf, A. (1981). Functional consequences of expanded aortic bulb: a model study. *Am. J. Physiol.* **240**, R200–10.

Camps, F. E. & Carpenter, R. G. (1972). *Sudden and Unexpected Deaths in Infancy (Cot Deaths)*. Bristol: John Wright & Sons, Ltd.

Castellini, M. A. & Somero, G. N. (1981). Buffering capacity of vertebrate muscle: correlations with potentials for anaerobic function. *J. Comp. Physiol.* **143**, 191–198.

Castellini, M. A., Somero, G. N. & Kooyman, G. L. (1981). Glycolytic enzyme activities in tissues of marine and terrestrial mammals. *Physiol. Zool.* **54**, 242–52.

Chernick, V. & Jansen, A. H. (1973). Initiation of respiratory movements in foetal sheep by hypoxic stimulation of foetal central chemoreceptor. In: *Foetal and Neonatal Physiology*, ed. R. S. Comline, K. W. Cross, G.

References

S. Dawes & P. W. Nathanielsz, pp. 213–16. Proceedings, Sir J. Barcroft Centenary Symposium, Cambridge University Press.

Chesler, A. & Himwich, H. E. (1944). Glycolysis in the parts of the central nervous system of cats and dogs during growth. *Am. J. Physiol.* **142**, 544–9.

Chien, S. (1967). Role of the sympathetic nervous system in hemorrhage. *Physiol. Rev.* **47**, 214–88.

Christensen, E. H. & Dill, D. B. (1935). Oxygen dissociation curves of bird blood. *J. Biol. Chem.* **109**, 443–8.

Clausen, G. (1964). Bradycardia during a submersion in the water vole, *Arvicola terrestris. Arbok Univ. Bergen* **17**, 3–6.

Clausen, G. & Ersland, A. (1968). The respiratory properties of the blood of two diving rodents, the beaver and the water vole. *Respir. Physiol.* **5**, 350–9.

Cohen, M. V. (1972). Ventricular fibrillation precipitated by carotid sinus pressure: case report and review of the literature. *Am. Heart J.* **84**, 681–6.

Cohn, H. E. & Moses, M. (1966). The critical interval in cadaver kidney transplantation: Function of canine renal autotransplants after variable periods of ischemia. *Surgery* **60**, 750–3.

Cohn, J. E., Krog, J. & Shannon, R. (1968). Cardiopulmonary responses to head immersion in domestic geese. *J. Appl. Physiol.* **25**, 36–41.

Colebatch, H. J. H. & Halmagyi, D. F. J. (1962). Reflex airway reaction to fluid aspiration. *J. Appl. Physiol.* **17**, 787–94.

Comline, R. S. & Silver, M. (1975). Placental transfer of blood gases. *Brit. Med. Bull.* **31**, 25–31.

Cordero, L., Jr & Hon, E. H. (1971). Neonatal bradycardia following nasopharyngeal stimulation. *J. Pediat.* **78**, 441–7.

Cordier, D. & Heymans, C. (1935). Le Centre Respiratoire. Paris. *Ann. Physiol. Physiochem. Biol.* **11**, 535–757.

Corriol, J. H. & Rohner, J. J. (1968a). New facts about bradycardia in breath-holding divers. *R. Subaq. Phys. Hyperb. Med.* **1**, 24–7.

Corriol, J. & Rohner, J. J. (1968b). Rôle de la température de l'eau dans la bradycardie d'immersion de la face. *Arch. Sci. Physiol.* **22**, 265–74.

Corriol, J., Rohner, J. J. & Fondarai, J. (1966). Origine et signification de la bradycardie chez le plongeur en apnê. *Path. et Biol.* **14**, 1185–91.

Cot, C. (1931). *Les Axphyxies Accidentelles (Submersion, Electrocution, Intoxication Oxycarbonique)*. Etude Clinique, Thérapeutique et Préventive. Paris: Editions Médicales N. Maloine.

Craig, A. B. (1961a). Underwater swimming and loss of consciousness. *J. Am. Med. Assn.* **176**, 255–8.

Craig, A. B. (1961b). Causes of loss of consciousness during underwater swimming. *J. Appl. Physiol.* **16**, 583–6.

Craig, A. B. (1963). Heart rate responses to apneic underwater diving and to breath-holding in man. *J. Appl. Physiol.* **18**, 854–62.

Crew, A. (1933). *The Old Bailey*, pp. 159–76. London: Nicholson & Watson, Ltd.

Cross, E. R. (1965). Taravana, diving syndrome in the Tuamotu diver. In:

References

Physiology of Breath-hold Diving and the Ama of Japan, ed. H. Rahn & T. Yokoyama, p. 207–19, publ. 1341. Washington, DC: Natl. Acad. Sci., Natl. Res. Council.

Cross, K. W. (1966). Resuscitation of the asphyxiated infant. *Brit. Med. Bull.* **22**, 73–8.

Cross, K. W. & Lewis, S. R. (1971). Upper respiratory obstruction and cot death. *Arch. Dis. Child.* **46**, 211–13.

Cross, K. W., Hooper, J. M. D. & Oppe, T. E. (1953). The effect of inhalation of carbon dioxide in air on the respiration of the full-term and premature infant. *J. Physiol.* **122**, 264–73.

Crowe, J. H. & Cooper, A. F., Jr (1971). Cryptobiosis. *Sci. Am.* **224**, 30–6.

Cunningham, D. J. C., Hey, E. N., Patrick, J. M. & Lloyd, B. B. (1963). The effect of noradrenaline infusion on the relation between pulmonary ventilation and the alveolar pO_2 and pCO_2 in man. *Anns. N. Y. Acad. Sci.* **109**, 756–70.

Daly, M. de B. (1972). Interaction of cardiovascular reflexes. *The Scientific Basis of Medicine Annual Reviews 1972*, pp. 307–32.

Daly, M. de B. & Angell-James, J. E. (1975). Role of the arterial chemoreceptors in the control of the cardiovascular responses to breath-hold in diving. In: *The Peripheral Arterial Chemoreceptors*, ed. J. Purves, pp. 387–407. Cambridge University Press.

Daly, M. de B. & Angell-James, J. E. (1979). The diving response and its possible clinical implications. *Internat. Med.* **1**, 12–19.

Daly, M. de B. & Hazzledine, J. L. (1963). The effects of artificially induced hyperventilation on the primary cardiac reflex response to stimulation of the carotid bodies in the dog. *J. Physiol.* **168**, 872–89.

Daly, M. de B. & Robinson, B. H. (1968). An analysis of the reflex systemic vasodilator response elicited by lung inflation in the dog. *J. Physiol.* **195**, 387–406.

Daly, M. de B. & Scott, M. J. (1958). The effects of stimulation of the carotid body chemoreceptors on heart rate in the dog. *J. Physiol.* **144**, 148–66.

Daly, M. de B. & Scott, M. J. (1963). The cardiovascular responses to stimulation of the carotid body chemoreceptors in the dog. *J. Physiol.* **165**, 179–97.

Daly, M. de B. & Scott, M. J. (1964). The cardiovascular effects of hypoxia in the dog with special reference to the contribution of the carotid body chemoreceptor. *J. Physiol.* **173**, 201–14.

Daly, M. de B. & Taton, A. (1979). Upper airways reflexes evoked by bronchodilator drugs administered in pressurized aerosol form in the conscious rabbit. *IRCS Medical Science* **7**, 255.

Daly, M. de B. & Ungar, A. (1966). Comparison of the reflex responses elicited by stimulation of the separately perfused carotid and aortic body chemoreceptors in the dog. *J. Physiol.* **182**, 379–403.

Daly, M. de B., Hazzledine, J. L. & Ungar, A. (1967). The reflex effects of alterations in lung volume on systemic vascular resistance in the dog. *J. Physiol.* **188**, 331–51.

137

References

Daly, M. de B., Elsner, R. & Angell-James, J. E. (1977). Cardiorespiratory control by carotid chemoreceptors during experimental dives in the seal. *Am. J. Physiol.* **232**, H508–16.

Daly, M. de B., Korner, P. I., Angell-James, J. E. & Oliver, J. R. (1978). Cardiovascular–respiratory reflex interactions between carotid bodies and upper-airways receptors in the monkey. *Am. J. Physiol.* **234**, H293–9.

Daly, M. de B., Angell-James, J. E. & Elsner, R. (1979a). Role of carotid-body chemoreceptors and their reflex interactions in bradycardia and cardiac arrest. *Lancet* **1** (No. 8119), 764–7.

Daly, M. de B., Angell-James, J. E. & Elsner, R. (1979b). The diving response, possible clinical implications. *Practioner, Special Report: Drowning* 19–23.

Daly, M. de B., Angell-James, J. E. & Elsner, R. (1980). Cardiovascular–respiratory interactions in breath-hold diving. In: *Central Interaction Between Respiratory and Cardiovascular Control Systems*, ed. H. P. Koepchen, S. M. Hilton & A. Trzebski, pp. 224–31. Satellite Symposium of the XXVIIth International Congress of Physiological Sciences. Berlin: Springer-Verlag.

Dauber, I. M., Krauss, A. N., Symchyck, P. S. & Auld, P. A. M. (1976). Renal failure following perinatal anoxia. *J. Pediat.* **88**, 851–5.

Davenport, H. W. (1974). *The ABC of Acid–Base Chemistry*, 6th edn, 124 pp. Chicago: University of Chicago Press.

Davis, F. M., Charlier, R., Saumarez, R. & Muller, V. (1972). Some physiological responses to the stress of aqualung diving. *Aerospace Med.* **43**, 1083–8.

Dawes, G. S. (1965). Oxygen supply and consumption in late fetal life, and the onset of breathing at birth. In: *Handbook of Physiology*, sect. 3, *Respiration*, vol, II, ed. W. O. Fenn & H. Rahn, pp. 1313–28. Washington DC: A. Physiol. Soc.

Dawes, G. S. (1968). *Foetal and Neonatal Physiology*. Chicago: Year Book Medical Publishers, Inc.

Dawes, G. S. & Mott, J. C. (1964). Changes in O_2 distribution and consumption in foetal lambs with variations in umbilical blood flow. *J. Physiol.* **170**, 524–40.

Dawes, G. S., Mott, J. C. & Shelley, H. J. (1959). The importance of cardiac glycogen for the maintenance of life in foetal lambs and newborn animals during anoxia. *J. Physiol.* **146**, 516–38.

Dawes, G. S., Jacobson, H. N., Mott, J. C., Shelley, J. H. & Stafford, A. (1963). The treatment of asphyxiated mature foetal lambs and rhesus monkeys with intravenous glucose and sodium carbonate. *J. Physiol.* **169**, 167–84.

Dawes, G. S., Lewis, B. V., Milligan, J. E., Roach, M. R. & Talner, N. S. (1968). Vasomotor responses in the hind limbs of foetal and new-born lambs to asphyxia and aortic chemoreceptor stimulation. *J. Physiol.* **195**, 55–81.

Dawes, G. S., Fox, H. E. & Richards, R. T. (1972). Variations in asphyxial

138

gasping with fetal age in lambs and guinea pigs. *Quart. J. Exper. Physiol.* **57**, 131–8.

Dejours, P. (1975). *Principles of Comparative Respiratory Physiology*, pp. 18–19. Amsterdam: North Holland Publishing Co.

Denison, D. M. & Kooyman, G. L. (1973). The structure and function of the small airways in pinniped and sea otter lungs. *Respir. Physiol.* **17**, 1–10.

De Villota, E. D., Barat, G., Peral, P., Juffé, A., Fernandez de Miguel, J. M. & Avello, F. (1973). Recovery from profound hypothermia with cardiac arrest after immersion. *Brit. Med.* **4**, 394–5.

Dixon, W. E. & Brodie, T. G. (1903). Contributions to the physiology of the lungs. Part I. The bronchial muscles, their innervation and the action of drugs upon them. *J. Physiol.* **29**, 97–173.

Dormer, K. J., Denn, M. J. & Stone, H. L. (1977). Cerebral blood flow in the sea lion (*Zalophus californianus*) during voluntary dives. *Comp. Biochem. Physiol.* **58A**, 11–18.

Downing, S. E. & Lee, J. C. (1975). Laryngeal chemosensitivity: a possible mechanism for sudden infant death. *Pediatrics* **55**, 640–9.

Drabek, C. M. (1975). Some anatomical aspects of the cardiovascular system of antarctic seals and their possible function significance in diving. *J. Morphol.* **145**, 85–106.

Drabek, C. M. (1977). Some anatomical and functional aspects of seal hearts and aortea. In: *Functional Anatomy of Marine Mammals*, ed. R. J. Harrison, vol. 3, pp. 217–34. London: Academic Press.

Drummond, P. C. & Jones, D. R. (1979). The initiation and maintenance of bradycardia in a diving mammal, the muskrat, *Ondatra zibethica*. *J. Physiol.* **290**, 253–71.

DuBois, A. B. & Marshall, R. (1957). Measurement of pulmonary capillary blood flow and gas exchange throughout the respiratory cycle in man. *J. Clin. Invest.* **36**, 1566–71.

Dumitru, A. P. & Hamilton, F. G. (1963). A mechanism of drowning. *Anesth. Analg.* **42**, 170–5.

Dutz, H. & Kretzschmar, G. (1954). Die Veranderungen in der Funktion beider Nieren nach einseitiger vollstandiger Ischamie. *Z. Exptl. Med.* **123**, 497–510.

Dykes, R. W. (1974). Factors related to the dive reflex in harbor seals: sensory contributions from the trigeminal region. *Can. J. Physiol. Pharmacol.* **52**, 259–65.

Ebbecke, U. (1944). Ubersichten der Gesichtsreflex des Trigeminus als Wärmeschutzreflex (Wind-und Wetterreflex) des Kopfes. *Klin, Wschr.* **23**, 141–5.

Ebbecke, U. & Knüchel, F. (1943). Uber den Trigeminus-Atem-Schluck- und Herzreflex beim Kaninchen. *Pflügers Arch. Ges. Physiol.* **247**, 255–63.

Editorial (1973). The tourniquet, instrument or weapon? *Can. Med. Assn. J.* **109**, 827.

Edwards, H. A., Benstead, J. G., Brown, K., Makary, A. Z. & Menon, N.

References

K. (1970). Apparent death with accidental hypothermia. *Brit. J. Anaesth.* **42**, 906–8.

Eldridge, F. L. (1972). The importance of timing on the respiratory effects of intermittent carotid body chemoreceptor stimulation. *J. Physiol.* **222**, 319–33.

Elsner, R. (1965). Heart rate response in forced versus trained experimental dives in pinnipeds. *Hvalrådets Skrifter, Norske Videnskaps-Akad., Oslo* **48**, 24–9.

Elsner, R. (1966). Diving bradycardia in the unrestrained hippopotamus. *Nature* **212**, 408.

Elsner, R. (1969). Cardiovascular adjustments to diving. In: *The Biology of Marine Mammals*, ed. H. T. Andersen, pp. 117–45. New York: Academic Press.

Elsner, R. (1978). Asphyxial survival: diving seals and fetal sheep. In: *Fetal and Newborn Cardiovascular Physiology*, ed. L. D. Longo, pp. 399–411. New York: Garland Publishing.

Elsner, R. (1979). Diving capability in seals – learned or innate? *Abstracts, 3rd Conf. Biol. Mar. Mammals*, p. 16. Seattle.

Elsner, R. & Gooden, B. A. (1970). Reduction of reactive hyperemia in the human forearm by face immersion. *J. Appl. Physiol.* **29**, 627–30.

Elsner, R. W., Garey, W. F. & Scholander, P. F. (1963). Selective ischemia in diving man. *Am. Heart J.* **65**, 571–3.

Elsner, R., Franklin, D. L. & Van Citters, R. L. (1964a). Cardiac output during diving in an unrestrained sea lion. *Nature* **202**, 809–910.

Elsner, R., Scholander, P. F., Craig, A. B., Dimond, E. G., Irving, L., Pilson, M., Johansen, K. & Bradstreet, E. (1964b). A venous oxygen reservoir in the diving elephant seal. *Physiologist* **7**, 124.

Elsner, R., Franklin, D. L., Van Citters, R. L. & Kenney, D. W. (1966a). Cardiovascular defense against asphyxia. *Science* **153**, 941–9.

Elsner, R., Kenney, D. W. & Burgess, K. (1966b). Diving bradycardia in the trained dolphin. *Nature* **212**, 407–8.

Elsner, R., Kooyman, G. L. & Drabek, C. M. (1969). Diving duration in pregnant Weddell seals. In: *Antarctic Ecology*, ed. M. Holdgate, pp. 477–82. New York: Academic Press.

Elsner, R., Hammond, D. D. & Parker, H. R. (1970a). Circulatory responses to asphyxia in pregnant and fetal animals: a comparative study of Weddell seals and sheep. *Yale J. Biol. Med.* **42**, 202–17.

Elsner, R., Shurley, J. T., Hammond, D. D. & Brooks, R. E. (1970b). Cerebral tolerance to hypoxemia in asphyxiated Weddell seals. *Respir. Physiol.* **9**, 287–97.

Elsner, R., Gooden, B. A. & Robinson, S. M. (1971a). Arterial blood gas changes and the diving response in man. *Aust. J. Exp. Biol. Med. Sci.* **49**, 435–44.

Elsner, R., Gooden, B. A. & Robinson, S. M. (1971b). *Circulatory effects of human face immersion: Chemoreceptor influences.* Preprints Congr. Physiol. Sci. Satellite Symp. Recent Progr. in the Fundamental Physiology of Diving, 25th, pp. 12–13. Marseille.

Elsner, R. W., Hanafee, W. N. & Hammond, D. D. (1971c). Angiography of the inferior vena cava of the harbor seal during simulated diving. *Am. J. Physiol.* **220**, 1155–7.

Elsner, R., Hammel, H. T. & Heller, H. C. (1975). Combined thermal and diving stresses in the harbor seal *Phoca vitulina*: A preliminary report. *Rapp. P.-v. Réun. Cons. Int. Explor. Mer.* **169**, 437–40.

Elsner, R., Franklin, D., White, F., McKown, D. & Kemper, S. (1976). Responses of harbor seal heart to ischemia and hypoxia. *Circulation* **54**, supp. II: 68.

Elsner, R., Angell-James, J. E. & Daly, M. de B. (1977a). Carotid body chemoreceptor reflexes and their interactions in the seal. *Am. J. Physiol.* **232**, H517–25.

Elsner, R., Hammond, D. D., Denison, D. M. & Wyburn, R. (1977b). Temperature regulation in the newborn Weddell seal *Leptonychotes weddelli*. In: *Adaptations Within Antarctic Ecosystems*, ed. G. A. Llano, pp. 531–40. Houston, Texas: Gulf Publishing.

Elsner, R., Blix, A. S. & Kjekshus, J. K. (1978). Tissue perfusion and ischemia in diving seals. *Physiologist* **21**, 33.

Elsner, R., Blix, A. S., Franklin, D., Kjekshus, J., Mueller, G. & Millard, R. W. (1981a). Coronary blood flow and *in vitro* vascular smooth muscle oscillations during diving and hypoxia in the seal. *Fed. Proc.* **40**, 446.

Elsner, R., Millard, R., Kjekshus, J., Blix, A., Hol, R. & Sordahl, L. (1981b). Cardiac adaptations in diving seals. In: Adv. Physiol. Sci., vol. 20, *Advances in Animal and Comparative Physiology*, ed. G. Pethes & V. L. Frenyo, pp. 269–74. 28th Inter. Cong. Physiol. Sci. Oxford: Pergamon Press.

Fainer, D. C., Martin, C. G. & Ivy, A. C. (1951). Resuscitation of dogs from fresh water drowning. *J. Appl. Physiol.* **3**, 417–26.

Fairbanks, E. S. & Kilgore, D. L., Jr (1978). Post-dive oxygen consumption of restrained and unrestrained muskrats (*Ondatra zibethica*). *Comp. Biochem. Physiol.* **59A**, 113–17.

Fales, J. T., Heisey, S. R. & Zierler, K. L. (1962). Blood flow and oxygen uptake by muscle during and after partial venous occlusion. *Am. J. Physiol.* **203**, 470–4.

Fazekas, J. F. & Bessman, A. N. (1953). Coma mechanisms. *Am. J. Med.* **15**, 804–12.

Feigl, E. & Folkow, B. (1963). Cardiovascular responses in 'diving' and during brain stem stimulation in ducks. *Acta Physiol. Scand.* **57**, 99–110.

Feinstein, R., Pinsker, H., Schmale, M. & Gooden, B. A. (1977). Bradycardial response in aplysia exposed to air. *J. Comp. Physiol.* **122**, 311–24.

Fenn, W. O. (1965). Greetings from the International Union of Physiological Sciences. In: *Physiology of Breath-hold Diving and the Ama of Japan*, ed. H. Rahn & T. Yokoyama, p. 3, publ, 1341. Washington, DC: Natl. Acad. Sci., Natl. Res. Council.

Ferrante, F. L. & Opdyke, D. F. (1969). Mammalian ventricular function during submersion asphyxia. *J. Appl. Physiol.* **26**, 561–70.

Ferren, H. & Elsner, R. (1979). Diving physiology of the ringed seal: adaptations and implications. *Proc. 29th Alaska Sci. Conf.*, pp. 379–87.

References

Finley, J. P., Bonet, J. F. & Waxman, M. B. (1979). Autonomic pathways responsible for bradycardia on facial immersion. *J. Appl. Physiol.* **47**, 1218–22.

Fitzhardinge, P. M. (1977). Complications of asphyxia and their therapy. In: *Intrauterine Asphyxia and the Developing Fetal Brain*, ed. L. Gluck, pp. 285–92. Chicago: Year Book Medical Publishers, Inc.

Flatt, A. E. (1972). Tourniquet time in hand surgery. *Arch. Surg.* **104**, 190–2.

Folinsbee, L. (1974). Cardiovascular response to apneic immersion in cool and warm water. *J. Appl. Physiol.* **36**, 226–32.

Folkow, B. & Rubinstein, E. H. (1965). Effect of brain stimulation on 'diving' in ducks. *Hvalrådets Skrifter, Norske Videnskaps-Akad., Oslo* **48**, 30–41.

Folkow, B. & Yonce, L. R. (1967). The negative intropic effect of vagal stimulation on the heart ventricles of the duck. *Acta Physiol. Scand.* **71**, 77–84.

Folkow, B., Fuxe, K. & Sonnenschein, R. R. (1966). Responses of skeletal musculature and its vasculature during 'diving' in the duck: pecularities of the adrenergic vasoconstrictor innervation. *Acta Physiol. Scand.* **67**, 327–42.

Forster, R. P. & Nyboer, J. (1955). Effect of induced apnea on cardiovascular renal functions in the rabbit. *Am. J. Physiol.* **183**, 149–54.

Frankel, H. L., Mathias, C. J. & Spalding, J. M. (1975). Mechanisms of reflex cardiac arrest in tetroplegic patients. *Lancet* **2**, (7946) 1183–5.

Franklin, D. L., Schlegel, W. L. & Watson, N. W. (1963). Ultrasonic Doppler shift blood flow meter: circuitry and practical applications. In: *Biomedical Sciences Instrumentation*, pp. 390–15. New York: Plenum Press.

Franklin, K. J. (1951). Aspects of the circulation's economy. *Brit. Med. J.* **1**, 1343–9 or 1410–17.

Frantz, I. D., III, Adler, S. M., Thach, B. T. & Taensch, H. W., Jr (1976). Maturational effects on respiratory responses to carbon dioxide in premature infants. *J. Appl. Physiol.* **41**, 41–5.

French, J. W., Morgan, B. C. & Guntheroth, W. G. (1972). Infant monkeys: a model for crib death. *Am. J. Dis. Child* **123**, 480–4.

Frey, M. A. & Kenney, R. A. (1974). Systolic time intervals during face immersion bradycardia. *Fed. Proc.* **33**, 327.

Friedman, S. M., Johnson, R. L. & Friedman, C. L. (1954). The pattern of recovery of renal function following renal artery occlusion in the dog. *Circulation Res.* **2**, 231–5.

Friedman, W. F. & Kirkpatrick, S. E. (1977). Fetal cardiovascular adaptation to asphyxia. In: *Intrauterine Asphyxia and the Developing Fetal Brain*, ed. L. Gluck, pp. 149–65. Chicago: Year Book Medical Publishers, Inc.

Frishman, W. H. (1980). Clinical pharmacology of the new beta-adrenergic blocking drugs. Part 12, Beta-adrenoceptor blockade in myocardial infarction: the continuing controversy. *Am. Heart J.* **99**, 528–36.

References

Furnival, C. M., Linden, R. J. & Snow, H. M. (1973). The inotropic effect on the heart of stimulating the vagus in the dog, duck and toad. *J. Physiol.* **233**, 155–170.

Galantsev, V. P. (1977). Evolutionary adaptations of diving animals (in Russian) Leningrad, Acad. Sci. USSR, 1–192.

Gandevia, S. C., McCloskey, D. I. & Potter, E. K. (1978). Reflex bradycardia occurring in response to diving, nasopharyngeal stimulation and ocular pressure, and its modification by respiration and swallowing. *J. Physiol.* **276**, 383–94.

Gardner, E. (1942). Mechanism of certain forms of sudden death in medicolegal practice. *Med.-leg. Rev.* **10**, 120–33.

Garey, W. F. (1962). Cardiac responses of fishes in asphyxic environments. *Biol. Bull.* **122**, 362–8.

Gjone, R. (1961). Drukning som klinisk problem. *Norske Laegerforen* **81**, 418–20, 430.

Glaister, J. (1947). *Medical Jurisprudence and Toxicology*. Revised reprint of 8th edn. Baltimore: Williams and Wilkins Co.

Godfrey, S. & Campbell, E. J. M. (1968). The control of breath holding. *Respir. Physiol.* **5**, 385–400.

Goff, L. G. & Bartlett, R. G. (1957). Elevated end-tidal CO_2 in trained underwater swimmers. *J. Appl. Physiol.* **10**, 203–6.

Gooden, B. A. (1971a). Effects of face immersion on body temperature and tail blood flow in the rat. *Comp. Biochem. Physiol.* **40A**, 659–68.

Gooden, B. A. (1971b). The diving response in man, rat and echidna. MD thesis, Barr Smith Library, University of Adelaide, Australia.

Gooden, B. A. (1972). Drowning and the diving reflex in man. *Med. J. Aust.* **2**, 583–7.

Gooden, B. A. (1980a). A comparison *in vitro* of the vasoconstrictor responses of the mesenteric arterial vasculature from the chicken and the duckling to nervous stimulation and to noradrenaline. *Br. J. Pharmacol.* **68**, 263–73.

Gooden, B. A. (1980b). The effect of hypoxia on vasoconstrictor responses of isolated mesenteric arterial vasculature from chicken and duckling. *Comp. Biochem. Physiol.* **67C**, 219–22.

Gooden, B. A. (1982). The diving response in clinical medicine. *Aviat. Space Environ. Med.* **53**, 273–6.

Gooden, B. A., Lehman, R. G. & Pym, J. (1970). Role of the face in the cardiovascular responses to total immersion. *Aust. J. Ex. Biol. Med. Sci.* **48**, 687–90.

Gooden, B. A., Stone, H. L. & Young, S. (1974). Cardiac responses to snout immersion in trained dogs. *J. Physiol.* **242**, 405–14.

Gooden, B. A., Feinstein, R. & Skutt, H. R. (1975). Heart rate responses of scuba divers via ultrasonic telemetry. *Undersea Biomedical Research* **2**, 11–19.

Gooden, B. A., Holdstock, G. & Hampton, J. R. (1978). The magnitude of the bradycardia induced by face immersion in patients convalescing from myocardial infarction. *Cardiovascular Res.* **12**, 239–42.

Gratiolet, M. P. (1860). Recherches sur le système vasculaire sanguin de

References

l'hippopotame. *Compt. Rend. Acad. Sci.* **51**, 524–8.

Grausz, H., Amend, W. J. C. & Earley, L. E. (1971). Acute renal failure complicating submersion in sea water. *J. Am. Med. Assn.* **217**, 207–9.

Gregg, D. E. (1962). Hemodynamic factors in shock. In: *Shock, Pathogenesis and Therapy*, ed. K. D. Bock, pp. 50–60. New York: Academic Press.

Greiss, F. C. & Gobble, F. L. (1967). Effect of sympathetic nerve stimulation on the uterine vascular bed. *Am. J. Obstet. Gynecol.* **97**, 962–7.

Grenell, R. G. (1946). Central nervous system resistance: I. The effects of temporary arrest of the cerebral circulation for periods of 2 to 10 minutes. *J. Neuropath. Exper. Neurol.* **5**, 131–54.

Griffin, G. E. (1966). Near-Drowning: Its pathophysiology and treatment in man. *Milit. Med.* **131**, 12–21.

Guilleminault, C., Petaita, R., Souquet, M. & Dement, W. C. (1975). Apneas during sleep in infants: possible relationship with sudden infant death syndrome. *Science* **190**, 677–9.

Guilleminault, C., Tilkian, A., Lehrman, D., Forno, L. & Dement, W. C. (1977). Sleep apnea syndrome: states of sleep and autonomic dysfunction. *J. Neurol. Neurosurg. Psychiat.* **40**, 718–25.

Guyton, R. A. & Daggett, W. M. (1976). The evaluation of myocardial infarction: physiological basis for clinical intervention. In: *International Review of Physiology, Cardiovascular Physiology II*, ed. A. C. Cuyton & A. W. Cowley, pp. 305–40. Baltimore: University Park Press.

Halasz, N. A., Elsner, R., Garvie, R. S. & Grotke, G. T. (1974). Renal recovery from ischemia: a comparative study of seal and dog kidneys. *Am. J. Physiol.* **227**, 1331–5.

Haldane, J. S. (1922). *Respiration*, preface to the first edition. New Haven: Yale University Press.

Hamilton, W. F. & Mayo, J. P. (1944). Changes in the vital capacity when the body is immersed in water. *Am. J. Physiol.* **141**, 51–3.

Hammel, H. T., Elsner, R., Heller, H. C., Maggert, J. A. & Bainton, C. R. (1977). Thermoregulatory responses to altering hypothalamic temperature in the harbor seal. *Am. J. Physiol.* **232**, R18–26.

Hammond, D. D., Elsner, R., Simson, G. & Hubbard, R. (1969). Submersion bradycardia in the newborn elephant seal *Mirounga angustirostris. Am. J. Physiol.* **216**, 220–2.

Harding, P. E., Roman, D. & Whelan, R. F. (1965). Diving in man. *J. Physiol.* **181**, 401–9.

Harding, R., Johnson, P., Johnston, B. E., McClelland, M. F. & Wilkinson, A. R. (1976). Cardiovascular changes in new-born lambs during apnoea induced by stimulation of laryngeal receptors with water. *J. Physiol.* **256**, 35P–36P.

Harken, A. H. (1976). Hydrogen ion concentration and oxygen uptake in an isolated canine hindlimb. *J. Appl. Physiol.* **40**, 1–5.

Harrison, R. J. (Ed.) (1972). *Functional Anatomy of Marine Mammals*, vol. 1, preface, pp. vii–xv. London: Academic Press.

Harrison, R. J. & Kooyman, G. L. (1968). General physiology of the

144

pinnipedia. In: *The Behavior and Physiology of Pinnipeds*, ed. R. J. Harrison, R. C. Hubbard, R. S. Peterson, C. E. Rice & R. J. Schusterman, pp. 211–96. New York: Appleton-Century-Crofts.

Harrison, R. J. & Tomlinson, J. D. W. (1956). Observations on the venous system in certain pinnipedia and cetacea. *Zool. Soc. Lond., Proc.* **126**, 205–31.

Harrison, R. J. & Tomlinson, J. D. W. (1960). Normal and experimental diving in the common seal *Phoca vitulina*. *Mammalia* **24**, 386–99.

Harvey, W. (1628). *Exercitatio Anatomica de Motu Cordis et Sanguinis in Animalibus*, trans. by G. Whitteridge, 1976, p. 131. Oxford: Blackwell Scientific Publications.

Haymet, B. D. & McCloskey, D. E. (1975). Baroreceptor and chemoreceptor influences on heart rate during the respiratory cycle in the dog. *J. Physiol.* **245**, 699–712.

Heistad, D. D. & Wheeler, R. C. (1970). Simulated diving during hypoxia in man. *J. Appl. Physiol.* **38**, 652–6.

Heistad, D. D., Abboud, F. M. & Eckstein, J. W. (1968). Vasoconstrictor response to simulated diving in man. *J. Appl. Physiol.* **25**, 542–9.

Heller, H. C. & Colliver, G. W. (1974). CNS regulation of body temperature during hibernation. *Am. J. Physiol.* **227**, 583–9.

Heller, H. C. & Glotzbach, S. F. (1977). Thermoregulation during sleep and hibernation. *Int. Rev. Physiol,,* **15**, 147–88.

Heller, H. C. & Hammel, H. T. (1972). CNS control of body temperature during hibernation. *Comp. Biochem. Physiol.* **41A**, 349–60.

Heller, R., Mager, W. & von Schrötter, H. (1900). *Luftdruck Erkrankungen*. Wien: Alfred Hölder.

Hempleman, H. V. & Lockwood, A. P. M. (1978). *The Physiology of Diving in Man and other Animals*. The Institute of Biology, Studies in Biology No. 99, p. 1. London: Edward Arnold.

Henschel, J. M. & Coates, M. E. (1974). The toxicity of cow's milk to infant rabbits. *Proc. Nutr. Soc.* **33**, 112A.

Herber, F. J. (1948). Metabolic changes of blood and tissue gases during asphyxia. *Am. J. Physiol.* **152**, 687–95.

Hering, E. (1871). Uber den Einfluss der Atmung auf den Kreislauf. Zweite Mittheilung. Ubereine reflectorische Beziehung Zwischen Lunge und Herz. *S. B. Acad. Wiss. Wien.* **64**, 333–53.

Hill, L. (1912). *Caisson Sickness and the Physiology of Work in Compressed Air*. London: Edward Arnold.

Hillis, L. D. & Braunwald, E. (1977). Myocardial ischemia. *N. Eng. J. Med.* **296**, 971, 1034–41, 1093–6.

Hochachka, P. W. (1980). *Living Without Oxygen*, 181 pp. Cambridge, Mass.: Harvard University Press.

Hochachka, P. W. & Murphy, B. (1979). Metabolic status during diving and recovery in marine mammals. In: *Environmental Physiol. III*, vol. 20, ed. D. Robertshaw, pp. 253–87. Internat. Rev. Physiol. Baltimore: University Park Press.

Hochachka, P. W. & Storey, K. B. (1975). Metabolic consequences of diving in animals and man. *Science* **187**, 613–21.

145

References

Hol, R., Blix, A. S. & Myhre, H. O. (1975). Selective redistribution of the blood volume in the diving seal (*Pagophilus groenlandicus*). *Rapp. P.-v, Réun, Cons. Int. Explor. Mer.* **169**, 423–32.

Hollenberg, N. K. & Uvnäs, B. (1963). The role of the cardiovascular response in the resistance to asphyxia of avian divers. *Acta Physiol. Scand.* **58**, 150–61.

Holm, B. & Sørensen, S. C. (1972). The role of the carotid body in the diving reflex in the duck. *Respir. Physiol.* **15**, 302–9.

Holubár, J., Schück, M. & Saravec, C. (1952). Vliv nízké teploty na zotavení nervosvalov´ eho aparátu z úcinku ischemie. *Cas. lék. ces.* **91**, 755–61.

Hon, E. H. (1958). The electronic evaluation of the fetal heart rate. Prelim. Report. *Am. J. Obstet. Gynecol.* **75**, 1215–30.

Hon, E. H. (1966). The human fetal circulation in normal labor. In: *The Heart and Circulation in the Newborn and Infant*, ed. D. E. Cassels, pp. 37–52. New York: Grune & Stratton.

Hong, S. K. (1965). Hae-Nyo, the diving women of Korea. In: *Physiology of Breath-hold Diving and the Ama of Japan*, ed. H. Rahn & T. Yokoyama, pp. 99–111, publ. 1341. Washington, DC: Natl. Acad. Sci., Natl. Res. Council.

Hong, S. K. & Rahn, H. (1967). The diving women of Korea and Japan. *Sci. Am.* **216**, 34–43.

Hong, S. K., Song, S. H., Kim, P. K. & Suh, C. S. (1967). Seasonal observations on the cardiac rhythm during diving in the Korean Ama. *J. Appl. Physiol.* **23**, 18–22.

Hong, S. K., Moore, T. O., Seto, G., Park, H. K., Hiatt, W. R. & Bernauer, E. M. (1970). Lung volumes and apneic bradycardia in divers. *J. Appl. Physiol.* **29**, 172–6.

Hong, S. K., Lin, Y. C., Lally, D. A., Yim, B. J. B., Kominami, N., Hong, P. W. & Moore, T. O. (1971). Alveolar gas exchanges and cardiovascular functions during breath holding in air. *J. Appl. Physiol.* **30**, 540–7.

Hong, S. K., Moore, T. O., Lally, D. A. & Morlock, J. F. (1973). Heart rate response to apneic face immersion in hyperbaric heliox environment. *J. Appl. Physiol.* **34**, 770–4.

Hong, S. K., Ashwell-Erickson, S., Gigliotti, P., Mueller, G. & Elsner, R. (1981). Effects of N_2 and low pH on organic ion transport and electrolyte distribution in the harbor seal (*Phoca vitulina*) kidney slice. *Fed. Proc.* **40**, 475.

Hong, S. K., Ashwell-Erickson, S., Gigliotti, P. & Elsner, R. (1982). Effects of anoxia and low pH on organic ion transport and electrolyte distribution in the harbor seal (*Phoca virulina*) kidney slice. *J. Comp. Physiol. B*, in press.

Hornbein. T. F. (1968). The relation between stimulus to chemoreceptors and their response. In: *Arterial Chemoreceptors*, ed. R. W. Torrance, pp. 65–78. Oxford and Edinburgh: Blackwell Scientific Publications.

Huckabee, W. E. (1962). Uterine blood flow. *Am. J. Obstet. Gynecol.* **84**, 1623–33.

146

Hunt, N. G., Whitaker, D. K. & Willmott, N. J. (1975). Water temperature and the 'diving reflex'. *Lancet* 1, **572**.

Hunt, P. K. (1974). Effect and treatment of the 'diving reflex'. *Can. Med. Assn. J.* **111**, 1330–1.

Hunter, J. (1787). Observations on the structure and oeconomy of whales. *Phil. Trans.* **77**, 371–450.

Huntley, A. C., Costa, D. P., Walker, J. A. & Berger, R. J. (1981). Cessation of respiration during sleep: a mode of energy conservation in the elephant seal. *Abstracts, 4th Conf. Biol. Mar. Mammals*, p. 60. San Francisco.

Huxley, F. M. (1912). Reflex apnoea and resistance to asphyxia in the duck. MD Thesis, University of Manchester.

Huxley, F. M. (1913*a*). On the reflex nature of apnoea in the duck in diving. I. The reflex nature of submersion apnoea. *Quart. J. Exptl Physiol.* **6**, 147–57.

Huxley, F. M. (1913*b*). On the reflex nature of apnoea in the duck in diving. II. Reflex postural apnoea. *Quart. J. Exptl Physiol.* **6**, 159–82.

Huxley, F. M. (1913*c*). On the resistance to asphyxia of the duck in diving. *Quart. J. Exptl Physiol.* **6**, 183–96.

Imbach, P., Kabus, K. & Tönz, O. (1975). Erfolgreiche Behandlung eines schweren Ertrinkungsunfalls nach 20 minütiger Submersion. *Schweiz. Med. Wschr.* **105**, 1605–11.

Irving, L. (1934). On the ability of warm blooded animals to survive without breathing. *Sci. Mon. N.Y.* **38**, 422–8.

Irving, L. (1938). Changes in the blood flow through the brain and muscles during the arrest of breathing. *Am. J. Physiol.* **122**, 207–14.

Irving, L. (1939). Respiration in diving mammals. *Physiol. Rev.* **19**, 112–34.

Irving, L. (1963). Bradycardia in human divers. *J. Appl. Physiol.* **18**, 489–91.

Irving, L. (1964). Comparative anatomy and physiology of gas transport mechanisms. In: *Handbook of Physiol.*, sect. 3, *Respiration*, vol. I, ed. W. O. Fenn & H. Rahn, pp. 177–212. Washington, DC: Am. Physiol. Soc.

Irving, L. (1969). Temperature regulation in marine mammals. In: *The Biology of Marine Mammals*, ed. H. T. Anderson, pp. 147–74. New York: Academic Press.

Irving, L., Solandt, O. M., Solandt, D. Y. & Fisher, K. C. (1935). The respiratory metabolism of the seal and its adjustment to diving. *J. Cell. Comp. Physiol.* **7**, 137–51.

Irving, L., Scholander, P. F. & Grinnell, S. W. (1941*a*). The respiration of the porpoise, *Tursiops truncatus*. *J. Cell. Comp. Physiol.* **17**, 145–68.

Irving, L., Scholander, P. F. & Grinnell, S. W. (1941*b*). Significance of the heart rate to the diving ability of seals. *J. Cell. Comp. Physiol.* **18**, 283–97.

Irving, L., Scholander, P. F. & Grinnell, S. W. (1942*a*). The regulation of arterial blood pressure in the seal during diving. *Am. J. Physiol.* **135**, 557–66.

Irving, L., Scholander, P. F. & Grinnell, S. W. (1942*b*). Experimental studies of the respiration of sloths. *J. Cell. Comp. Physiol.* **20**, 189–210.

Irving, L., Peyton, L. J. & Monson, M. (1956). Metabolism and insulation of swine as bare-skinned mammals. *J. Appl. Physiol.* **9**, 421–6.

Irving, L., Peyton, L. J., Bahn, C. H. & Peterson, R. S. (1963). Action of the heart and breathing during development of fur seals. *Physiol. Zool.* **36**, 1–20.

Jackson, D. C. (1968). Metabolic depression and oxygen depletion in the diving turtle. *J. Appl. Physiol.* **24**, 503–9.

Jackson, D. C. & Schmidt-Nielsen, K. (1966). Heat production during diving in the fresh water turtle (*Pseudemys scripta*). *J. Cell. Physiol.* **67**, 225–32.

James, L. S., Weisbrot, I. M., Prince, C. E., Holaday, D. A. & Apgar, V. (1958). The acid–base status of human infants in relation to birth asphyxia and the onset of respiration. *J. Pediat.* **52**, 379–94.

Jarrell, S. O. (1979). Characteristics of lactate dehydrogenase and hexokinase from harbor seal (*Phoca vitulina richardsi*) and domestic dog (*Canis familiaris*) cerebrum. MS thesis, University of Alaska.

Joels, N. & Samueloff, M. (1956). The activity of the medullary centres in diffusion respiration. *J. Physiol.* **133**, 360–72.

Johansen, K. (1959). Heart activity during experimental diving of snakes. *Am. J. Physiol.* **197**, 604–6.

Johansen, K. (1964). Regional distribution of circulating blood during submersion asphyxia in the duck. *Acta Physiol. Scand.* **62**, 1–9.

Johansen, K. & Aakhus, T. (1963). Central cardiovascular responses to submersion asphyxia in the duck. *Acta. Physiol. Scand.* **62**, 1–9.

Johansen, K., Lenfant, C. & Grigg, G. C. (1966). Respiratory properties of blood and responses to diving of the platypus *Ornithorhynchus anatinus* (Shaw). *Comp. Biochem. Physiol.* **18**, 597–608.

Johlin, J. M. & Moreland, F. B. (1933). Studies of the blood picture of the turtle after complete anoxia. *J. Biol. Chem.* **103**, 107–114.

Johnson, P., Dawes, G. S. & Robinson, J. S. (1972). Maintenance of breathing in newborn lamb. *Arch. Dis. Child.* **47**, 151.

Johnson, P., Robinson, J. S. & Salisbury, D. (1973). The onset and control of breathing after birth. In: *Foetal and Neonatal Physiology*, ed. R. S. Comline, K. W. Cross, G. S. Dawes & P. W. Nathanilsz, pp. 217–21. Proceedings, Sir J. Bancroft Centenary Symposium. Cambridge University Press.

Johnson, P., Salisbury, D. M. & Storey, A. T. (1975). Apnoea induced by stimulation of sensory receptors in the larynx. In: *Symposium on Development of Upper Respiratory Anatomy and Function. Implications for Sudden Infant Death Syndrome*, ed. J. Bosma. Washington, DC: US Dept. Health, Education and Welfare.

Johnson, R. H. & Spalding, J. M. K. (1974). *Disorders of the Autonomic Nervous System.* Oxford: Blackwell.

Jones, D. R. (1973). Systemic arterial baroreceptors in ducks and the consequences of their denervation on some cardiovascular responses to diving. *J. Physiol.* **234**, 499–518.

References

Jones, D. R. (1976). The control of breathing in birds with particular reference to the initiation and maintenance of diving apnea. *Fed. Proc.* **35**, 1975–82.

Jones, D. R. & Johansen, K. (1972). The blood vascular system of birds. In: *Avian Biology*, ed. D. S. Farner & J. R. King, vol. 2, pp. 157–285. New York: Academic Press.

Jones, D. R. & Purves, M. J. (1970). The carotid body in the duck and the consequences of its denervation upon the cardiac responses to immersion. *J. Physiol.* **211**, 279–94.

Jones, D. R. & West, N. H. (1978). The contribution of arterial chemoreceptors and baroreceptors to diving reflexes in birds. In: *Respiratory Function in Birds, Adult and Embryonic*, ed. J. Piiper, pp. 95–103. Berlin: Springer-Verlag.

Jones, D. R., Fisher, H. D., McTaggart, S. & West, N. H. (1973). Heart rate during breath-holding and diving in the unrestrained harbor seal (*Phoca vitulina richardsi*). *Can. J. Zool.* **51**, 671–80.

Jones, D. R., Bryan, R. M., West, N. H., Lord, R. H. & Clark, B. (1979). Regional distribution of blood flow during diving in the duck (*Anas platyrhynchos*). *Can. J. Zool.* **57**, 995–1002.

Jones, D. R., Milsom, W. K. & West, N. H. (1980). Cardiac receptors in ducks; the effect of their stimulation and blockade on diving bradycardia. *Am. J. Physiol.* **238**, R50–6.

Kabat, H. (1940). The greater resistance of very young animals to arrest of the brain circulation. *Am. J. Physiol.* **130**, 588–99.

Kabat, H., Dennis, C. & Baker, A. B. (1941). Recovery of function following arrest of the brain circulation. *Am. J. Physiol.* **132**, 737–47.

Kagen, L. J. (1973). *Myoglobin: Biochemical, Physiological and Clinical Aspects.* 147 pp. New York: Columbia University Press.

Kagen, L. J. & Christian, C. L. (1966). Immunologic measurements of myoglobin in human adult and fetal skeletal muscle. *Am. J. Physiol.* **211**, 656–60.

Kanwisher, J. W., Gabrielsen, G. & Kanwisher, N. (1981). Free and forced dives in birds. *Science* **211**, 717–19.

Katz, A. M. (1977). *Physiology of the Heart*, 450 pp. New York: Raven Press.

Kawakami, Y., Natelson, B. H. & DuBois, A. (1967). Cardiovascular effects of face immersion and factors affecting diving reflex in man. *J. Appl. Physiol.* **23**, 964–70.

Keatinge, W. R. (1969). *Survival in Cold Water*, 131 pp. Oxford: Blackwell Scientific Publications.

Kerem, D. & Elsner, R. (1973a). Cerebral tolerance to asphyxial hypoxia in the dog. *Am. J. Physiol.* **225**, 593–600.

Kerem, D. & Elsner, R. (1973b). Cerebral tolerance to asphyxial hypoxia in the harbor seal. *Respir. Physiol.* **19**, 188–200.

Kerem, D. & Salzano, J. (1974). Effect of high ambient pressure on human apneic bradycardia. *J. Appl. Physiol.* **37**, 108–11.

Kerem, D., Hammond, D. D. & Elsner, R. (1973). Tissue glycogen levels

149

in the Weddell seal: A possible adaptation to asphyxial hypoxia. *Comp. Biochem. Physiol.* **45A**, 731–6.

King, R. B. & Webster, I. W. (1964). A case of recovery from drowning and prolonged anoxia. *Med. J. Aust.* **1**, 919–20.

Kirk, E. S. & Honig, C. R. (1964). Nonuniform distribution of blood flow and gradients of oxygen tension within the heart. *Am. J. Physiol.* **207**, 661–8.

Kita, H. (1965). Review of activities: Harvest, seasons and diving patterns. In: *Physiology of Breath-hold Diving and the Ama of Japan*, ed. H. Rahn & T. Yokoyama, pp. 41–55, publ. 1341. Washington, DC: Natl. Acad. Sci., Natl. Res. Council.

Kjekshus, J. K. (1973). Mechanism for flow distribution in normal and ischemic myocardium during increased ventricular preload in the dog. *Circulation Res.* **33**, 489–99.

Kjekshus, J. K. & Blix, A. S. (1977). How seals avoid myocardial infarction during diving. *Scand. J. Clin. Lab. Invest.* **37**, 95–8.

Kjekshus, J. K., Maroko, P. K. & Sobel, B. E. (1972). Distribution of myocardial injury and its relation to epicardial ST-segment changes after coronary artery occlusion in the dog. *Cardiovascular Res.* **6**, 490–9.

Kjekshus, J. K., Blix, A. S., Grottum, P. & Aasen, A. O. (1981a). Beneficial effects of vagal stimulation on the ischemic myocardium during betareceptor blockade. *Scand. J. Clin. Lab. Invest.* **41**, 383–90.

Kjekshus, J. K., Blix, A. S., Elsner, R., Millard, R. & Hol, R. (1981b). The multifactorial approach to myocardial salvage, the experience from diving seals. *Acta Med. Scand.* **210**, supp. 651, 49–57.

Kjekshus, J. K., Blix, A. S., Elsner, R., Hol, R. & Amundsen, E. (1982). Myocardial blood flow and metabolism in the diving seal. *Am. J. Physiol.* **242**, R79–104.

Kjellmer, I. (1965). On the competition between metabolic vasodilation and neurogenic vasoconstriction in skeletal muscle. *Acta Physiol. Scand.* **63**, 450–9.

Kobayasi, S. & Ogawa, T. (1973). Effect of water temperature during nonapneic facial immersion in man. *Jap. J. Physiol.* **23**, 613–24.

Kobinger, W. & Oda, M. (1969). Effects of sympathetic blocking substances on the diving reflex of ducks. *Europ. J. Pharmacol.* **7**, 289–95.

Kodama, A. M., Elsner, R. & Pace, N. (1977). Effects of growth, diving history and high altitude on blood oxygen in harbor seals. *J. Appl. Physiol.* **42**, 852–8.

Kooyman, G. L. (1966). Maximum diving capacities of the Weddell seal, *Leptonychotes weddelli. Science* **151**, 1553–4.

Kooyman, G. L. (1968). An analysis of some behavioral and physiological characteristics related to diving in the Weddell seal. In: *Biology of the Antarctic Seas III*, ed. W. L. Schmitt & G. A. Llano, pp. 227–61. Washington, DC: Am. Geophys. Union.

Kooyman, G. L. (1972). Deep diving behavior and effects of pressure in reptiles, birds and mammals. *Soc. Exp. Biol. Symp.* **26**, 295–306.

Kooyman, G. L. (1975). Physiology of freely diving Weddell seals. *Rapp. P.-v. Réun. Cons. Int. Explor. Mer.* **169**, 441–4.

References

Kooyman, G. L. & Campbell, W. B. (1972). Heart rates in freely diving Weddell seals *Leptonychotes weddelli*. *Comp. Biochem. Physiol.* **43A**, 31–6.

Kooyman, G. L., Kerem, D. H., Campbell, W. B. & Wright, J. J. (1971). Pulmonary function in freely diving Weddell seals *Leptonychotes weddelli*. *Respir. Physiol.* **12**, 271–82.

Kooyman, G. L., Schroeder, J. P., Denison, D. M., Hammond, D. D., Wright, J. J. & Bergman, W. P. (1972). Blood nitrogen tensions of seals during simulated deep dives. *Am. J. Physiol.* **223**, 1016–20.

Kooyman, G. L., Kerem, D. H., Campbell, W. B. & Wright, J. J. (1973). Pulmonary gas exchange in freely diving Weddell seals *Leptonychotes weddelli*. *Respir. Physiol.* **17**, 283–90.

Kooyman, G. L., Wahrenbrock, E. A., Castellini, M. A., Davis, R. W. & Sinnett, E. E. (1980). Aerobic and anaerobic metabolism during voluntary diving in Weddell seals: evidence of preferred pathways from blood chemistry and behavior. *J. Comp. Physiol.* **138**, 335–46.

Kooyman, G. L., Castellini, M. A. & Davis, R. W. (1981). Physiology of diving in marine mammals. *Ann. Rev. Physiol.* **43**, 343–56.

Koppyanyi, T. & Dooley, M. S. (1929). Submergence and postural apnea in the muskrat. *Am. J. Physiol.* **88**, 592–5.

Kordy, M. T., Neil, E. & Palmer, J. F. (1975). The influence of laryngeal afferent stimulation on cardiac vagal responses to carotid chemoreceptor excitation. *J. Physiol.* **247**, 24–5.

Koschier, F. J., Elsner, R., Holleman, D. F. & Hong, S. K. (1978). Organic anion transport by renal cortical slices of harbor seals, *Phoca vitulina*. *Comp. Biochem. Physiol.* **60A**, 289–92.

Kountz, W. B. (1936). Revival of human hearts, *Anns. Int. Med.* **10**, 330–6.

Kratschmer, F. (1870). Uber Reflexe von der Nasenschleimhaut auf Athmung und Kreislauf. *Sber. Akad. Wiss. Wien* **62**, 147–70.

Kvittingen, T. D. & Naess, A. (1963). Recovery from drowning in fresh water. *Brit. Med. J.* **1**, 1315–17.

Ladner, C. C., Brinkman, R., III, Weston, P. & Assali, N. S. (1970). Dynamics of uterine circulation in pregnant and nonpregnant sheep. *Am. J. Physiol.* **218**, 257–63.

Lamb, L. E., Dermskian, G. & Sarnoff, C. A. (1958). Significant cardiac arrhythmias induced by common respiratory maneuvers. *Am. J. Cardiol.* **2**, 563–71.

Lampson, R. S., Schaeffer, W. C. & Lincoln, J. R. (1948). Acute circulatory arrest from ventricular fibrillation for twenty-seven minutes, with complete recovery. *J. Am. Med. Assn.* **137**, 1575–8.

Langlois, P. & Richet, C. (1898). Des gaz expirés par les canards plongés dans l'eau. *Compt. Rend. Séanc. Soc. Biol.* **5**, 483–6.

Langman, V. R., Bandinette, R. V. & Taylor, C. R. (1981). Maximum aerobic capacity of wild and domestic canids compared. *Fed. Proc.* **40**, 432.

La Veck, G. D. (1972). Sudden Infant Death Syndrome. Selected annotated bibliography. 1960–1971. DHEW Publ. NO. (NIH) 73–237. US Dept of Health, Education and Welfare.

References

Leape, L. L., Holder, T. M., Franklin, J. D., Amoury, R. A. & Ashcraft, K. W. (1977). Respiratory arrest in infants secondary to gastroesophageal reflux. *Pediatrics*, **60**, 924–8.

Lees, A. D. (1955). *The Physiology of Diapause in Arthropods*, 151 pp. Cambridge University Press.

Le Gallois, M. (1812). *Expériences sur le Principe de la Vie*. Paris: Hautel.

Leivestad, H. (1960). The effect of prolonged submersion on the metabolism and the heart rate in the toad. *Arbok Univ. Bergen, Mat.-Nat. Ser.* **5**, 1–15.

Leivestad, H., Andersen, H. & Scholander, P. F. (1957). Physiological response to air exposure in the codfish. *Science* **126**, 505.

Lekven, J., Mjøs, O. D. & Kjekshus, J. K. (1973). Compensatory mechanisms during graded myocardial ischemia. *Am. J. Cardiol.* **31**, 467–73.

Lenfant, C. (1969). Physiological properties of blood of marine mammals. In: *The Biology of Marine Mammals*, ed. H. T. Andersen, pp. 95–116. New York: Academic Press.

Lenfant, C., Elsner, R., Kooyman, G. L. & Drabek, C. M. (1969*a*). Respiratory function of the blood of the adult and fetal Weddell seal *Leptonychotes weddelli*. *Am. J. Physiol.* **216**, 1595–7.

Lenfant, C., Elsner, R., Kooyman, G. L. & Drabek, C. M. (1969*b*). Tolerance to sustained hypoxia in the Weddell seal *Leptonychotes weddelli*. In: *Antarctic Ecology*, ed. M. W. Holdgate, pp. 471–6. New York: Academic Press.

Lenfant, C., Kooyman, G. L., Elsner, R. & Drabek, C. M. (1969*c*). Respiratory function of blood of the Adélie penguin *Pygoscelis adeliae*. *Am. J. Physiol.* **216**, 1598–600.

Lenfant, C., Johansen, K. & Torrance, J. D. (1970). Gas transport and oxygen storage capacity in some pinnipeds and the sea otter. *Respir. Physiol.* **9**, 277–86.

Liggins, G. C., France, J. T., Knox, B. S. & Zapol, W. M. (1979). High corticosteroid levels in plasma of adult and foetal Weddell seals (*Leptonychotes weddelli*). *Acta Endocrinol.* **90**, 718–26.

Liggins, G. C., Qvist, J., Hochachka, P. W., Murphy, B. J., Creasy, R. K., Schneider, R. C., Snider, M. T. & Zapol, W. M. (1980). Fetal cardiovascular and metabolic responses to simulated diving in the Weddell seal. *J. Appl. Physiol.* **49**, 424–30.

Lin, Y. C. & Baker, D. G. (1975). Cardiac output and its distribution during diving in the rat. *Am. J. Physiol.* **228**, 733–7.

Lombroso, U. (1913). Uber die Reflexhemmung des Herzens wahrend der reflektorischen Atmungshemmung bei verschiedenen Tieren. *Z. Biol.* **61**, 517–38.

Lopes, O. W. & Palmer, J. F. (1976). Proposed respiratory 'gating' mechanism for cardiac slowing. *Nature* **264**, 454–6.

Lowrence, P. B., Nickel, J. F., Smythe, C. M. C. & Bradley, S. E. (1956). Comparison of the effect of anoxic anoxia and apnea on renal function in the harbor seal (*Phoca vitulina* L.) *J. Cell. Comp. Physiol.* **48**, 35–42.

de Lumley, H. (1969). A paleolithic camp at Nice. *Sci. Am.* **220**, 42–50.

References

Lutz, P. L., LaManna, J. C., Adams, M. R. & Rosenthal, M. (1980). Cerebral resistance to anoxia in the marine turtle. *Respir. Physiol.* **41**, 241–51.

MacLeod, R. D. M. & Scott, M. J. (1964). The heart rate responses to carotid body chemoreceptor stimulation in the cat. *J. Physiol.* **175**, 193–202.

Marshall, S. B., Owens, J. C. & Swan, H. (1956). Temporary circulatory occlusion to the brain of the hypothermic dog. *Arch. Surg.* **72**, 98–106.

Maseri, A., L'Abbate, A., Baroldi, G., Chierchia, S., Marzilli, M., Bellestra, A. M., Severi, S., Parodi, O., Biagini, A., Distante, A. & Pesola, A. (1978). Coronary vasospasm as a possible cause of myocardial infarction. *N. Eng. J. Med.* **299**, 1271–7.

McKean, T. & Carlton, C. (1977). Oxygen storage in beavers. *J. Appl. Physiol.* **42**, 545–7.

McWhirter, N. (1980). *Guinness Book of Records*, 27th edn, pp. 24, 212. London: Guinness.

McWhirter, N. & McWhirter, R. (1973). *Guinness Book of World Records*, p. 45. New York: Bantam.

Mendez-Bauer, C., Poseiro, J. J., Arellano-Hernandez, Q., Zambrana, M. A. & Caldeyro-Barcia, R. (1963). Effects of atropine on the heart rate of the human fetus during labor. *Am. J. Obstet. Gynecol.* **85**, 1033–53.

Menick, F. J., White, F. C. & Bloor, C. M. (1971). Coronary collateral circulation: determination of an anatomical anastomotic index of functional flow capacity. *Am. Heart J.* **82**, 503–10.

Meyer, J. S., Gotah, F., Tazaki, T., Hamaguchi, K., Ishikawa, S., Nouailhat, F. & Symond, L. (1962). Regional cerebral blood flow and metabolism *in vitro*. *Arch. Neurol.* **7**, 560–78.

Michenfelder, J. D. & Theye, R. A. (1968). Hypothermia: Effect on canine brain and whole-body metabolism. *Anesthesiology* **29**, 1107–12.

Michenfelder, J. D., Terry, H. R., Daw, E. F., MacCarty, C. S. & Uihlein, A. (1963). Profound hypothermia in neurosurgery: open-chest versus closed-chest techniques. *Anesthesiology* **24**, 177–84.

Miles, S. (1962). *Underwater Medicine*. London: Staples Press.

Millard, R. W., Johansen, K. & Milsom, W. K. (1973). Radiotelemetry of cardiovascular responses to exercise and diving in penguins. *Comp. Biochem. Physiol.* **46A**, 227–40.

Millard, R. W., Kjekshus, J., Blix, A. S., Franklin, D. & Elsner, R. (1980). Coronary vasoconstriction in diving seals: a natural model of vasospasm. *Fed. Proc.* **39**, 398.

Millen, J. E., Murdaugh, H. V., Jr, Bauer, C. B. & Robin, E. D. (1964). Circulatory adaptation to diving in the freshwater turtle. *Science* **145**, 591–3.

Mithoefer, J. C. (1965). The breaking point of breath-holding. In: *Physiology of Breath-hold Diving and the Ama of Japan*, ed. H. Rahn & T. Yokoyama, pp. 195–205, publ. 1341. Washington, DC: Natl. Acad. Sci., Natl. Res. Council.

Modell, J. H. (1971). *The Pathophysiology and Treatment of Drowning and Near-Drowning*. Springfield, Illinois: Charles C. Thomas, Publisher.

References

Modell, J. H., Calderwood, H. W., Ruiz, B. C., Downs, J. B. & Chapman, R. (1974). Effect of ventilatory patterns on arterial oxygenation after near-drowning in sea water. *Anesthesiology* **40**, 376–84.

Mohri, H., Dillard, D. H., Crawford, E. W., Martin, W. E. & Merendino, K. A. (1969). Method of surface-induced deep hypothermia for open-heart surgery in infants. *J. Thorac, Cardiovasc. Surg.* **58**, 262–70.

Mohrman, D. E. & Feigl, E. O. (1978). Competition between sympathetic vasoconstriction and metabolic vasodilation in the canine coronary circulation. *Circulation Res.* **42**, 79–86.

Moore, T. O., Lin, Y. C., Lally, D. A. & Hong, S. K. (1972a). Effects of temperature, immersion and ambient pressure on human apneic bradycardia. *J. Appl. Physiol.* **33**, 36–41.

Moore, T. O., Elsner, R., Lally, D. A. & Hong, S. K. (1972b). The effects of alveolar PO_2 and PCO_2 on apneic bradycardia in man. *J. Appl. Physiol,* **34**, 795–98.

Moritz, A. R. (1944). Chemical methods of the determination of death by drowning. *Physiol. Rev.* **24**, 70–88.

Mueller, H. S., Ayres, S. M., Religa, A. & Evans, R. G. (1974). Propranolol in the treatment of acute myocardial infarction. *Circulation* **49**, 1078–87.

Murdaugh, H. V., Jr & Jackson, J. E. (1962). Heart rate and blood lactic acid concentration during experimental diving of water snakes. *Am. J. Physiol.* **202**, 1163–5.

Murdaugh, H. V., Schmidt-Nielsen, B., Wood, J. W. & Mitchell, W. L. (1961a). Cessation of renal function during diving in the trained seal (*Phoca vitulina*). *J. Cell. Comp. Physiol.* **58**, 261–5.

Murdaugh, H. V., Jr, Seabury, J. C. & Mitchell, W. L. (1961b). Electrocardiogram of the diving seal. *Circulation Res.* **9**, 358–61.

Murdaugh, H. V., Robin, E. D., Millen, J. E., Drewry, W. F. & Weiss, E. (1966). Adaptations to diving in the harbor seal: cardiac output during diving. *Am. J. Physiol.* **210**, 176–80.

Murdaugh, H. V., Cross, C. E., Millen, J. E., Gee, J. B. L. & Robin, E. D. (1968). Dissociation of bradycardia and arterial constriction during diving in the seal *Phoca vitulina*. *Science* **162**, 364–5.

Murie, J. (1874). Research upon the anatomy of the pinnipedia, pt III. Descriptive anatomy of the sea lion (*Otaria jubata*). *Trans. Zool. Soc. Lond.* **8**, 501–82.

Murphy, B., Zapol, W. M. & Hochachka, P. W. (1980). Metabolic activities of the heart, lung, and brain during diving and recovery in the Weddell seal. *J. Appl. Physiol.* **48**, 596–605.

Myers, R. W., Pearlman, A. P., Hyman, R. M., Goldstein, R. A., Kent, K. M., Goldstein, R. E. & Epstein, S. E. (1974). Beneficial effects of vagal stimulation and bradycardia during experimental acute myocardial ischemia. *Circulation* **49**, 943–7.

Naeye, R. L., Fisher, R., Ryser, M. & Whalen, P. (1976). Carotid body in the sudden infant death syndrome. *Science* **191**, 567–9.

Nagel, E. L., Morgane, P. J., McFarland, W. L. & Galliano, R. E. (1968). Rete mirabile of dolphin: its pressure-damping effect on cerebral circulation. *Science* **161**, 898–900.

154

Nakamura, H. (1921). The oxygen use of muscle and the effect of sympathetic nerves on it. *J. Physiol.* **55**, 100–10.

Neely, J. R. & Morgan, H. E. (1974). Relationship between carbohydrate and lipid metabolism and the energy balance of heart muscle. *Ann. Rev. Physiol.* **36**, 413–59.

Nemiroff, M. J. (1977). Accidental cold water immersion and survival characteristics. *Undersea Biomed. Res.* **4**, A56.

Neubuerger, K. T. (1954). Lesions of the human brain following circulatory arrest. *J. Neuropath. Exper. Neurol.* **8**, 144–60.

Newell, R. C. (1973). Factors affecting the respiration of intertidal invertebrates. *Amer. Zool.* **13**, 513–28.

Noble, A. B. (1946). Cerebral anoxia complicating spinal anaesthesia. *Can. Med. Assn. J.* **54**, 378–9.

Nukada, M. (1965). Historical development of the Ama's diving activities. In: *Physiology of Breath-hold Diving and the Ama of Japan*, ed. H. Rahn & T. Yokoyama, pp. 25–40, publ. 1341. Washington, DC: Natl. Acad. Sci., Natl. Res. Council.

Ogura, J. H. & Lam, R. L. (1953). Anatomical and physiological correlations on stimulating the human superior laryngeal nerve. *Laryngoscope* **63**, 947–59.

Ohlsson, K. & Beckman, M. (1964). Drowning – reflections based on two cases. *Acta Chir. Scand.* **128**, 327–39.

Olsen, C. R., Fanestil, D. D. & Scholander, P. F. (1962a). Some effects of breath-holding and apneic underwater diving on cardiac rhythm in man. *J. Appl. Physiol.* **17**, 461–6.

Olsen, C. R., Fanestil, D. D. & Scholander, P. F. (1962b). Some effects of apneic underwater diving on blood gases, lactate and pressure in man. *J. Appl. Physiol.* **17**, 938–42.

Olsen, C. R., Hale, F. C. & Elsner, R. (1969). Mechanics of ventilation in the pilot whale. *Respir. Physiol.* **7**, 137–49.

Packer, B. S., Altman, M., Cross, C. E., Murdaugh, H. V., Jr, Linta, J. M. & Robin, E. D. (1969). Adaptations to diving in the harbor seal: oxygen stores and supply. *Am. J. Physiol.* **217**, 903–9.

Paletta, F. X., Willman, V & Ship, A. G. (1960). Prolonged tourniquet ischemia of extremities. *J. Bone Joint Surg.* **42A**, 945–50.

Pantridge, J. F., Webb, S. W. & Adgey, A. A. (1975). Autonomic disturbance at the onset of acute myocardial infarction. In: *National Conf. on Stnds. for Cardiopulmonary Resuscitation and Emergency Cardiac Care*, pp. 229–33. Dallas: Am. Heart Assoc.

Pappenheimer, J. R. (1941). Blood flow, arterial oxygen saturation and oxygen consumption in the isolated perfused hind limb of the dog. *J. Physiol.* **99**, 283–303.

Parfrey, P. & Sheehan, J. D. (1975). Individual facial areas in the human circulatory response to immersion. *Irish J. Med. Sci.* **144**, 335–42.

Parish, W. E., Barrett, A. M., Combs, R. A., Gunther, M. & Camps, F. E. (1960). Hypersensitivity to milk and sudden death in infancy. *Lancet* **2**, 1106–10.

Parkes, A. (1973). Ischemic effects of external and internal pressure on the upper limb. *Hand* **5**, 105–12.

References

Påsche, A. (1976). Hypoxia in freely diving hooded seal, *Cystophora cristata*. *Comp. Biochem. Physiol.* **55A**, 319–22.

Paulev, P. (1968*a*). Cardiac rhythm during breath-holding and water immersion in man. *Acta Physiol. Scand.* **73**, 138–50.

Paulev, P. (1968*b*). Impaired consciousness during breath-hold diving and breath-holding in air. *Revue de Physiologie Subaquatique et Medicine Hyperbare* **1**, 16–20.

Pearn, J. H., Bart, R. D. & Yamaoka, R. (1979). Neurologic sequelae after childhood near-drowning: a total population study from Hawaii. *Pediatrics* **64**, 187–91.

Phillips, R. A. & Hamilton, P. B. (1948). Effect of 20, 60, and 120 minutes of renal ischemia on glomerular and tubular function. *Am. J. Physiol.* **152**, 523–30.

Pickering, T. & Bolton-Maggs, P. (1975). Treatment of paroxysmal supraventricular tachycardia. Letter to the *Lancet* **1**, 340.

Pickwell, G. V. (1968). Energy metabolism in ducks during submergence asphyxia: assessment by a direct method. *Comp. Biochem. Physiol.* **27**, 455–85.

Preston, J. B., McFadden, S. & Moe, G. K. (1959). Atrioventricular transmission in young mammals. *Am. J. Physiol.* **197**, 236–40.

Raffucci, F. L. & Wangensteen, O. H. (1951). Tolerance of dogs to occlusion of entire afferent vascular inflow to the liver. *Surgical Forum* **1**, 191–5.

Raffucci, F. L., Lewis, F. J. & Wangensteen, O. H. (1953). Hypothermia in experimental hepatic surgery. *Proc, Soc. Exper. Biol. Med.* **83**, 639–40.

Rahn, H. (1964). Oxygen stores in man. In: *Oxygen in the Animal Organism*, ed. F. Dickens & E. Neil, vol. 31, pp. 609–19. International Biochemical Union symposium series. Oxford: Pergamon Press.

Rahn, H. (1965). G. Teruoka – his contribution to the physiology of the Ama. In: *Physiology of Breath-hold Diving and the Ama of Japan*, ed. H. Rahn & T. Yokoyama, pp. 9–10, publ. 1341. Washington, DC: Natl. Acad. Sci., Natl. Res. Council.

Rahn, H. & Yokoyama, T. (Eds.) (1965). *Physiology of Breath-hold Diving and the Ama of Japan*, publ. 1341. Washington, DC: Natl. Acad. Sci., Natl. Res. Council.

Raper, A. J., Richardson, D, W., Kontos, H. A. & Patterson, J. L., Jr (1967). Circulatory responses to breath-holding in man. *J. Appl. Physiol.* **22**, 201–6.

Redwood, D. R., Smith, E. R. & Epstein, S. E. (1972). Coronary artery occlusion in the conscious dog: effects of alterations in heart rate and arterial pressure on the degree of myocardial ischemia. *Circulation* **46**, 323–32.

Rein, H. & Schneider, M. (1937). Die lokale Stoffwechseleinschränkung bei reflektorisch – nervöser Durchblutungsdrosselung. *Pflügers Arch. Ges. Physiol.* **239**, 464–75.

Rennie, D. W., Covino, B. G., Howell, B. J., Song, S. H., Kang, B. S. & Hong, S. K. (1962). Physical insulation of Korean diving women. *J. Appl. Physiol.* **17**, 961–6.

156

Rhode, E. A., Peterson, T. M. & Elsner, R. (1977). Pressure–volume characteristics of the aorta of harbor seals, *Phoca vitulina. Fed. Proc.* **36**, 437.

Rhode, E. A., Elsner, R., Peterson, T. M., Campbell, K. B. & Spangler, W. (1983). Pressure–volume characteristics of the aortae of harbor and Weddell seals. Manuscript in preparation.

Richet, C. (1894). La résistance des canards à l'asphyxie. *Compt. Rend. Séanc. Soc. Biol.* **1**, 244–5.

Richet, C. (1899). De la résistance des canards à l'asphyxie. *J. Physiol. Pathol. Gen.* **1**, 641–50.

Ridgway, S. H. (1972). Homeostasis in the aquatic environment. In: *Mammals of the Sea, Biology and Medicine*, ed. S. H. Ridgway, pp. 590–747. Springfield, Illinois: Charles C. Thomas, Publisher.

Ridgway, S. H., Scronce, B. L. & Kanwisher, J. (1969). Respiration and deep diving in the bottlenose porpoise. *Science* **166**, 1651–3.

Ridgway, S. H., Carder, D. A. & Clark, W. (1975). Conditioned bradycardia in the sea lion *Zalophus californianus. Nature* **256**, 37–8.

Rigatto, H. & Brady, J. P. (1972). Periodic breathing and apnea in preterm infants. In: *Evidence for hypo-ventilation possibly due to central respiratory depression. Pediatrics* **50**, 202–18.

Rigatto, H., Brady, J. P. & Verduzco, R. de la T. (1975). Chemoreceptor reflexes in preterm infants. 1. The effect of gestational and postnatal age on the ventilatory response to inhalation of 100% and 15% oxygen. *Pediatrics* **55**, 604–13.

Robin, E. D. (1966). Of seals and mitochondria. *N. Eng. J. Med.* **275**, 646–52.

Robin, E. D., Murdaugh, H. B., Jr, Pyron, W., Weiss, E. & Soteres, P. (1963). Adaptations to diving in the harbor seal – gas exchange and ventilatory response to CO_2. *Am. J. Physiol.* **205**, 1175–7.

Robin, E. D., Vester, J. W., Murdaugh, H. V., Jr & Millen, J. E. (1964). Prolonged anaerobiosis in a vertebrate: anaerobic metabolism in the freshwater turtle. *J. Cell. Comp. Physiol.* **63**, 287–97.

Robin, E. D., Ensinck, J., Hance, A. J., Newman, A., Lewiston, N., Cornell, L., Davis, R. W. & Theodore, J. (1981). Glucoregulation and simulated diving in the harbor seal *Phoca vitulina. Am. J. Physiol.* **241**, R293–300.

Robinson, D. (1939). The muscle hemoglobin of seals as an oxygen store in diving. *Science* **90**, 276–7.

Ronald, K., McCarter, R. & Selley, L. J. (1977). Venous circulation in the harp seal (*Papophilus groenlandicus*). In: *Functional Anatomy of Marine Mammals*, vol. 3, ed. R. J. Harrison, pp. 235–70. London: Academic Press.

Ross, D. N. (1957). Problems associated with the use of hypothermia in cardiac surgery. *Proc. Roy. Soc. Med.* **50**, 76–8.

Rossen, R., Kabat, H. & Anderson, J. P. (1943). Acute arrest of cerebral circulation in man. *Arch. Neurol. Psychiat.* **50**, 510–29.

Rovetto, M. J., Lamberton, W. F. & Neely, J. R. (1975). Mechanisms of glycolytic inhibition in ischemic rate hearts. *Circulation Res.* **37**, 742–51.

157

References

Rudolph, A. M. & Heymann, M. A. (1967). The circulation of the fetus in utero. Methods for studying distribution of blood flow, cardiac output and organ blood flow. *Circulation Res.* **21**, 163–84.

Rushmer, R. F., Van Citters, R. L. & Franklin, D. L. (1962). Shock: a semantic enigma. *Circulation*, **26**, 445–59.

Russetzki, J. (1925). Considérations sur les réflexes du nerf trijumeau sur le coeur. *Gaz. Hôp.* (Paris) **98**, 549.

Ryan, C., Hollenberg, M., Harvey, D. B. & Gwynn, R. (1976). Impaired parasympathetic responses in patients after myocardial infarction. *Am. J. Cardiol.* **37**, 1013–18.

Sasamoto, H. (1965). The electrocardiogram pattern of the diving Ama. In: *Physiology of Breath-hold Diving and the Ama of Japan*, ed. H. Rahn & T. Yokoyama, pp. 271–80, publ. 1341. Washington, DC: Natl. Acad. Sci., Natl. Res. Council.

Schaefer, K. E. (1955). The role of carbon dioxide in the physiology of human diving. In: *Proceedings of the Underwater Physiology Symposium.*, pp. 131–9, publ. 377. Washington, DC: Natl. Acad. Sci., Natl. Res. Council.

Schaefer, K. E. (1965). Adaptation to breath-hold diving. In: *Physiology of Breath-hold Diving and the Ama of Japan*, ed. H. Rahn & T. Yokoyama, pp. 237–52, publ. 1341. Washington, DC: Natl. Acad. Sci., Natl. Res. Council.

Schaefer, K. E. & Carey, C. R. (1962). Alveolar pathways during 90-foot breath-hold dives. *Science* **137**, 1051–2.

Scher, A. M. & Young, A. C. (1963). Servoanalysis of carotid sinus reflex effects on peripheral resistance. *Circulation Res.* **12**, 152–62.

Scherf, D. & Bornemann, C. (1966). Appearance of ventricular ectopic rhythm during carotid sinus pressure. *Dis. Chest* **50**, 530–2.

Scheuer, J. & Brachfeld, N. (1966). Coronary insufficiency: relations between hemodynamic electrical and biochemical parameters. *Circulation Res.* **18**, 178–9.

Schlosser, V. & Streicher, H. J. (1964). Studies on the length of time the heart will tolerate ischemia in normo- and hypothermia. *J. Thorac. Cardiovasc. Surg.* **48**, 430–5.

Schmidt, R. M., Kumada, M. & Sagawa, K. (1972). Cardiovascular responses to various pulsatile pressures in the carotid sinus. *Am. J. Physiol.* **223**, 1–7.

Schmidt-Nielsen, K., Taylor, C. R. & Shkolnik, A. (1971). Desert snails: problems of heat, water and food. *J. Exp. Biol.* **55**, 385–98.

Schneider, E. C. (1930). Observations on holding the breath. *Am. J. Physiol.* **94**, 464–70.

Scholander, P. F. (1940). Experimental investigations on the respiratory function in diving mammals and birds. *Hvalrådets Skrifter, Norske Videnskaps-Akad., Oslo* **22**, 1–131.

Scholander, P. F. (1960). Experimental studies on asphyxia in animals. In: *Oxygen Supply to the Foetus*, ed. J. Walker and A. C. Turnbull. Oxford: Blackwell.

Scholander, P. F. (1962). Physiological adaptations to diving in animals and man. *Harvey Lect.* **57**, 93–110.

Scholander, P. F. (1963). The master switch of life. *Sci. Am.* **209**, 92–106.

Scholander, P. F. (1964). Animals in aquatic environments: diving mammals and birds. In: *Handbook of Physiology*, sect. 4, *Adaptations to the Environment*, ed. D. B. Dill, E. F. Adolph & C. G. Wilber, pp. 729–39. Washington, DC: Am. Physiol. Soc.

Scholander, P. F. & Irving, L. (1941). Experimental investigations on the respiration and diving of the Florida Manatee. *J. Cell. Comp. Physiol.* **17**, 169–91.

Scholander, P. F., Irving, L. & Grinnell, S. W. (1942*a*). Aerobic and anaerobic changes in the seal muscles during diving. *J. Biol. Chem.* **142**, 431–40.

Scholander, P. F., Irving, L. & Grinnell, S. W. (1942*b*). On the temperature and metabolism of the seal during diving. *J. Cell. Comp. Physiol.* **19**, 67–78.

Scholander, P. F., Irving, L. & Grinnell, S. W. (1943). Respiration of the armadillo with possible implications as to its burrowing. *J. Cell. Comp. Physiol.* **21**, 53–63.

Scholander, P. F., Flagg, W., Hock, R. J. & Irving, L. (1953). Studies on the physiology of frozen plants and animals in the arctic. *J. Cell. Comp. Physiol.* **42**, 1–56.

Scholander, P. F., Hammel, H. T., LeMessurier, H., Hemmingsen, E. & Garey, W. (1962*a*). Circulatory adjustments in pearl divers. *J. Appl. Physiol.* **17**, 184–99.

Scholander, P. F., Bradstreet, E. & Garey, W. F. (1962*b*). Lactic acid response in the grunion. *Comp. Biochem. Physiol.* **6**, 201–3.

Sekar, T. S., MacDonnell, K. F., Namsirikul, P. & Herman, R. S. (1980). Survival after prolonged submersion in cold water without neurologic sequelae. Report of two cases. *Arch. Intern. Med.* **140**, 775–9.

Shannon, D. C., Marsland, D. W., Bould, J. B., Callahan, B., Todres, I. D. & Dennis, J. (1976). Central hypoventilation during quiet sleep in two infants. *Pediatrics* **57**, 342–6.

Shannon, D. C., Kelly, D. H. & O'Connel, K. (1977). Abnormal regulation of ventilation in infants at risk for sudden-infant-death syndrome. *N. Eng. J. Med.* **297**, 747–50.

Sharpey-Schafer, E. P. (1965). Effect of respiratory acts on the circulation. In: *Handbook of Physiology*, sect. 2, *Circulation*, vol. III, ed. W. F. Hamilton & P. Dow, pp. 1875–86. Washington, DC: *Am. Physiol. Soc.*

Shaw, E. B. (1968). Sudden unexpected death in infancy syndrome. *Am. J. Dis. Child.* **116**, 115–19.

Shaw, E. B. (1970). Sudden unexpected death in infancy. *Am. J. Dis. Child.* **119**, 416–18.

Shaw Wilgis, E. F. (1971). Observations on the effect of tourniquet ischemia. *J. Bone Joint Surg.* **53A**, 1343–6.

Shell, W. E. & Sobel, B. E. (1973). Deleterious effects of increased heart rate on infarct size in the conscious dog. *Am. J. Cardiol.* **31**, 474–9.

References

Siebke, H., Breivik, H., Rod, T. & Lind, B. (1975). Survival after 40 minutes' submersion without cerebral sequelae. *Lancet* **1**, 1275–7.

Simon, L. M., Robin, E. D., Elsner, R., Van Kessel, A. L. G. J. & Theodore, J. (1974). A biochemical basis for differences in maximal diving time in aquatic mammals. *Comp. Biochem. Physiol.* **47B**, 209–15.

Sinnett, E. E., Kooyman, G. L. & Wahrenbrock, E. A. (1978). Pulmonary circulation of the harbor seal. *J. Appl. Physiol.* **45**, 718–27.

Smith, H. (1930). Metabolism of the lungfish *Protopterus aethiopicus. J. Biol. Chem.* **88**, 97–130.

Smyth, H. S., Sleight, P. & Pickering, G. W. (1969). Reflex regulation of arterial pressure during sleep in man. A quantitative method of assessing baroreflex sensitivity. *Circulation Res.* **24**, 109–21.

Song, S. H., Kang, D. H., Kang, B. S. & Hong, S. K. (1963). Lung volumes and ventilatory responses to high CO_2 and low O_2 in the Ama. *J. Appl. Physiol.* **18**, 466–70.

Song, S. H., Lee, W. K., Chung, Y. A. & Hong, S. K. (1969). Mechanism of apneic bradycardia in man. *J. Appl. Physiol.* **27**, 323–7.

Sordahl, L. A., Steward, M. L., Mueller, G. & Elsner, R. (1981). Functional properties of seal heart mitochondria. *Fed. Proc.* **40**, 437.

Sordahl, L., Mueller, G. & Elsner, R. (1982). Comparative functional properties of mitochondria from seal and dog hearts. *J. Mol. Cell. Cardiol.*, in press.

South, F. E., Breazile, J. E., Dellman, J. D. & Epperly, A. D. (1969). Sleep, hibernation and hypothermia in the yellow-bellied marmot (*M. flaviventris*). In: *Depressed Metabolism*, ed. X. J. Musacchia & J. F. Saunders, pp. 277–310. New York: Elsevier.

Spencer, M. P., Gornall, T. A. & Poulter, T. C. (1967). Respiratory and cardiac activity of killer whales. *J. Appl. Physiol.* **22**, 974–81.

Stainsby, W. N. & Otis, A. B. (1964). Blood flow, blood oxygen tension, oxygen uptake and oxygen transport in skeletal muscle. *Am. J. Physiol.* **206**, 858–66.

Steinschneider, A. (1970). Possible cardiopulmonary mechanisms. In: *Sudden Infant Death Syndrome*, ed. B. Abraham, J. Bergman, B. Bechwith & C. Ray, pp. 181–98. Proc. of the Second Int. Conf. on Causes of Sudden Death in Infants. University of Washington Press: Seattle.

Steinschneider, A. (1972). Prolonged apnea and the sudden infant death syndrome. Clinical and laboratory observations. *Pediatrics* **50**, 646–54.

Strauss, M. B. (1970). Physiological aspects of mammalian breath-hold diving: a review. *Aerospace Med.* **41**, 1362–81.

Strφmme, S. B. & Blix, A. S. (1976). Indirect evidence for arterial chemoreceptor reflex facilitation by face immersion in man. *Aviat. Space Environ. Med.* **47**, 597–9.

Strφmme, S. B., Kerem, D. & Elsner, R. (1970). Diving bradycardia during rest and exercise and its relation to physical fitness. *J. Appl. Physiol.* **28**, 614–21.

Swann, H. G. (1956). Mechanism of circulatory failure in fresh and sea water drowning. *Circulation Res.* **4**, 241–4.

Szereda-Przestaszewska, M. & Widdicombe, J. G. (1973). Reflex effects of chemical irritation of the upper airways of the laryngeal lumen in cats. *Respir. Physiol.* **18**, 107–15.

Tanji, D. G., Weste, J. & Dykes, R. W. (1975). Interactions of respiration and the bradycardia of submersion in harbor seals. *Can. J. Physiol. Pharmacol.* **53**, 555–9.

Tatai, K. & Tatai, K. (1965). Anthropometric studies of the Japanese Ama. In: *Physiology of Breath-hold Diving and the Ama of Japan*, ed. H. Rahn & T. Yokoyama, p. 71–83, publ. 1341. Washington, DC: Natl. Acad. Sci., Natl. Res. Council.

Tazawa, H., Mikami, T. & Yoshimoto, C. (1971). Respiratory properties of chicken embryonic blood during development. *Respir. Physiol.* **13**, 160–70.

Tchobroutsky, C., Merlet, C. & Rey, P. (1969). The diving reflex in rabbit, sheep and newborn lamb and its afferent pathways. *Respir. Physiol.* **8**, 108–17.

Teruoka, G. (1932). Die Ama und ihre Arbeit. *Arbeitsphysiol.* **5**, 239–51.

Theroux, P., Franklin, D., Ross, J. & Kemper, W. S. (1974). Regional myocardial function during acute coronary artery occlusion and its modification by pharmacologic agents in the dog. *Circulation Res.* **35**, 896–908.

Tibes, V. & Stegemann, J. (1969). Das Verhalten der endexspiratorischen Atemgasdrucke, der O_2-Aufnahme and CO_2-Abgabe nach einfacher Apnoi im Wasser, an Land und apnoeischem Tauchen. *Pflügers Arch.* **311**, 300–11.

Turner, H. (1950). Case report: the mental state during recovery after heart arrest during anaesthesia. *J. Neurol. Neurosurg. Psychiat.* **13**, 153–5.

Valdes-Dapena, M. A. (1967). Sudden and unexpected death in infancy: a review of the world literature 1954–1966. *Pediatrics* **39**, 123–38.

Valdes-Dapena, M. (1970). Progress in SIDS Research, 1963–1969. In: *Sudden Infant Death Syndrome*, ed. A. B. Bergman, J. B. Blackwell & C. G. Ray, pp. 3–13. Proc of the Second Int. Conf. on Causes of Sudden Death in Infants. University of Washington Press: Seattle.

Van Citters, R. L., Franklin, D. L., Smith, O. A., Elsner, R. W. & Watson, N. W. (1965). Cardiovascular adaptations to diving in the northern elephant seal *Mirounga angustirostris*. *Comp. Biochem. Physiol.* **16**, 267–76.

Viamonte, M., Morgane, P. J., Galliano, R. E., Nagel, E. L. & McFarland, W. L. (1968). Angiography in the living dolphin and observations on blood supply to the brain. *Am. J. Physiol.* **214**, 1225–49.

Wahrenbrock, E. A., Maruschak, G. F., Elsner, R. & Kenney, D. W. (1974). Respiratory function and metabolism in two baleen whale calves. *Mar. Fish. Rev.* **36**, 3–9.

Walker, J. M., Paletta, F. X. & Cooper, T. (1959). Relationship of post-ischemic histopathological changes to muscle and subcuticular temperature patterns in the canine extremity. *Surgical Forum* **10**, 836–8.

Walker, J. M., Glotzbach, S. F., Berger, R. J. & Heller, H. C. (1977). Sleep and hibernation in ground squirrels (*Citellus* spp.): electrophysiological observations. *Am. J. Physiol.* **233**, R213–21.

References

Wallace, R. K., Benson, H. & Wilson, A. F. (1971). A wakeful hypo-metabolic physiological state. *Am. J. Physiol.* **221**, 795–9.

Warden, J. C. (1967). Respiratory insufficiency following near-drowning in sea water. *J. Am. Med. Assn.* **201**, 887–90.

Wasserman, K. & Mackenzie, A. (1957). Cardiac output in diving seals. *Bull. Tulane Univ. Med. Fac.* **16**, 105–10.

Watson, E. R. (1922). The case of Rex v. George Joseph Smith. *Trans. Med.-leg. Soc.* **16**, 134–64.

Wealthal, S. R. (1975). Factors resulting in a failure to interrupt apnoea. In: *Symposium on Development of Upper Respiratory Anatomy Are Function. Implication for Sudden Infant Death Syndrome*, ed. J. Bosma, pp. 212–32. Showacre. Washington, DC: US Department of Health, Education and Welfare.

Wedgwood, R. J. (1972). Review of USA experience, pp. 27–8. Conclusions by main speakers, pp. 98–99. In: *Sudden and Unexpected Deaths in Infancy (Cot Deaths)*, ed. F. E. Camps & R. G. Carpenter, Bristol: Wright & Son, Ltd.

Weiss, P. A. (1969). The living system: determinism stratified. In: *Beyond Reductionism, New Perspectives in the Life Sciences*, ed. A. Koestler & J. R. Smythies, p. 8. Boston: Beacon Press.

Whalen, W. J., Buerk, D. & Thuning, C. A. (1973). Blood flow–limited oxygen consumption in resting cat skeletal muscle. *Am. J. Physiol.* **224**, 763–8.

Whayne, T. F. & Killip, T. (1967). Simulated diving in man: comparison of facial stimuli and response in arrhythmia. *J. Appl. Physiol.* **22**, 800–7.

Whayne, T. F., Smith, N. T., Eger, E. I., Stoelting, R. K. & Whitcher, C. E. (1971). The effects of halothane anaesthesia on reflex cardiovascular responses to simulated diving and Valsalva manoeuvre. *Anesthesiology* **34**, 262–70.

White, F. N. (1981). A comparative physiological approach to hypothermia. *J. Thorac. Cardiovasc. Surg.* **82**, 821–831.

White, F. N. & Ross, G. (1966). Circulatory changes during experimental diving in the turtle. *Am. J. Physiol.* **211**, 15–18.

White, F. N., Ikeda, M. & Elsner, R. (1973). Adrenergic innervation of large arteries in the seal. *Comp. Gen. Pharmacol.* **4**, 271–6.

White, S. W. (1975). Central integration of the autonomic cardiorespiratory responses to nasopharyngeal stimulation in the rabbit. *Brain Research* **87**, 171–9.

White, S. W., McRitchie, R. J. & Franklin, D. L. (1974). Autonomic cardiovascular effects of nasal inhalation of cigarette smoke in the rabbit. *Aust. J. Biol. Med. Sci.* **52**, 111–26.

White, S., McRitchie, R. J. & Korner, P. I. (1975). Central nervous system control of cardiorespiratory nasopharyngeal reflexes in the rabbit. *Am. J. Physiol.* **228**, 404–9.

Whitney, R. J. (1953). The measurement of volume changes in human limbs. *J. Physiol.* **121**, 1–27.

Wildenthal, K. S., Leshin, S. J., Atkins, J. M. & Skelton, C. L. (1975). The diving reflex used to treat paroxysmal atrial tachycardia. *Lancet* **1**, 12–14.

Wolf, S. (1964). The bradycardia of the dive reflex – a possible mechanism of sudden death. *Trans. Am. Clin. Climat. Assn.* **76**, 192–200.

Wolf, S., Schneider, R. A. & Groover, M. E. (1965). Further studies on the circulatory and metabolic alterations of the oxygen-conserving (diving) reflex in man. *Trans. Assn. Am. Physns.* **78**, 242–54.

Wolff, W. I. (1950). Cardiac resuscitation. Complete recovery after 6 minutes of true circulatory arrest. *J. Am. Med. Soc.* **144**, 738–43.

Wood, J. E. (1965). *The Veins: Normal and Abnormal Function.* Boston: Little, Brown & Co.

Wood, S. C. & Johansen, K. (1974). Respiratory adaptations to diving in the Nile monitor lizard, *Varanus niloticus*. *J. Comp. Physiol.* **89**, 145–58.

Wyler, F. & Hof, R. (1977). Regional vascular responses to asphyxia in the rabbit. *Eu. J. Clin. Invest.* **7**, 67–70.

Wyss, V. (1956*a*). Electrocardiogramma di gosetti in apnea durante immersione in acqua a profondita diverse. *Boll. Soc. Ital., Biol. Sper.* **32**, 503–6.

Wyss, V. (1956*b*). Nuoto subacqueo in apnea e caratteri dell' electro-cardiogramma. *Boll. Soc. Ital., Biol. Sper.* **32**, 506–9.

Zapol, W. M., Liggins, G. C., Schneider, R. C., Qvist, J., Snider, M. T., Creasy, R. K. & Hochachka, P. W. (1979). Regional blood flow during simulated diving in the conscious Weddell seal. *J. Appl. Physiol.* **47**, 968–73.

INDEX

Index

carbon dioxide, 26, 28–9, 64, 125
  ventilatory stimulation in seals, 90
cardiac glycogen, 54, 111
cardiac output, 20–1
cardiac ventricular receptors, 85
cardiopulmonary resuscitation,
  119–20, 127
cardiovascular reactions, 13, 14–29,
  65–71
carotid body, 80
  isolated, 83
carotid sinus, 80
cat, 15
catecholamines, 65, 78
cerebral blood flow, 26
cerebral tolerance to asphyxia in seals,
  26, 36
chemoreceptor reflexes, 80–5, 88, 90,
  99–101
circulation, 48–9
  in diving seals, 7
  muscle, 16, 27
cold exposure,
  Ama divers, 62
  experimental subjects, 95–6
  in near-drowning, 116–19
  in seals, 3
contractility, ventricular, 39
control mechanisms, neural
  in human subjects, 92–103
  in natural divers, 75–92
coronary blood flow
  in dogs, 43
  in seals, 39, 41
coronary vasoconstriction, 41–2
cryptobiosis, 9

death, sudden underwater, 120–2
decerebration and diving ducks, 77
decompression sickness, 3–4
dives, aerobic, 5, 13, 16
diving depth, 3, 97
diving duration
  free, 4, 5, 13, 17, 75, 91–2
  maximum, 2, 3, 63
  restrained, 17, 75
  trained, 4, 17
diving experience, 102
'diving reflex', misnamed, 7
diving response, 4, 5, 7
  clinical applications, 126–7
  development in juvenile seals, 34,
  36

dog, 15, 20, 23, 24, 25, 89
dolphin, 4, 17
dormancy, 8
duck, 17, 75
  arteries, reactions to stimulation,
  45–6
  control mechanisms during diving
  75, 76–7, 80–1, 84–5, 86–7
  metabolic rate during dives, 12
  resistance to asphyxia, 14–15
dugong, 16

echidna, 20, 25–6
electrocardiogram, 17, 20, 39, 41, 67
electroencephalogram of seals, 36, 113
emotion, effect on diving response, 4,
  91–2, 101, 102
enzymes
  cardiac, 39
  glycolytic, 56
exercise and diving, 50, 52, 57, 100

face immersion
  clinical applications, 126–7
  human, 68, 92–6
fear and diving response, 91–2, 102
fetal asphyxia, 104–14
fetal breathing, 114
fish out of water, 8
flowmeter, 20, 23–6, 41, 107, 109
fright, effect on seal's heart rate, 4

glucagon, 56
glucose, blood, 54, 56, 106, 107
glycerol, 57
glycogen, 111
  in seal tissues, 54

haematocrit, 36, 59
haemoglobin, 3, 35, 59
  fetal, 105
heart
  relative intolerance to asphyxia, 30,
  32, 38–43
  muscle, ischaemia in, 38
heart rate, 15, 16–20, 17, 18, 19, 43
heat production of diving ducks, toads
  and turtles, 12
hibernation, 8, 10–11, 36
hippopotamus, bradycardia during
  free dives, 17
history
  diving physiology, 14–15, 76–7, 78

166